MAMMAL REMAINS FROM ARCHÆOLOGICAL SITES

PART I

SOUTHEASTERN AND SOUTHWESTERN UNITED STATES

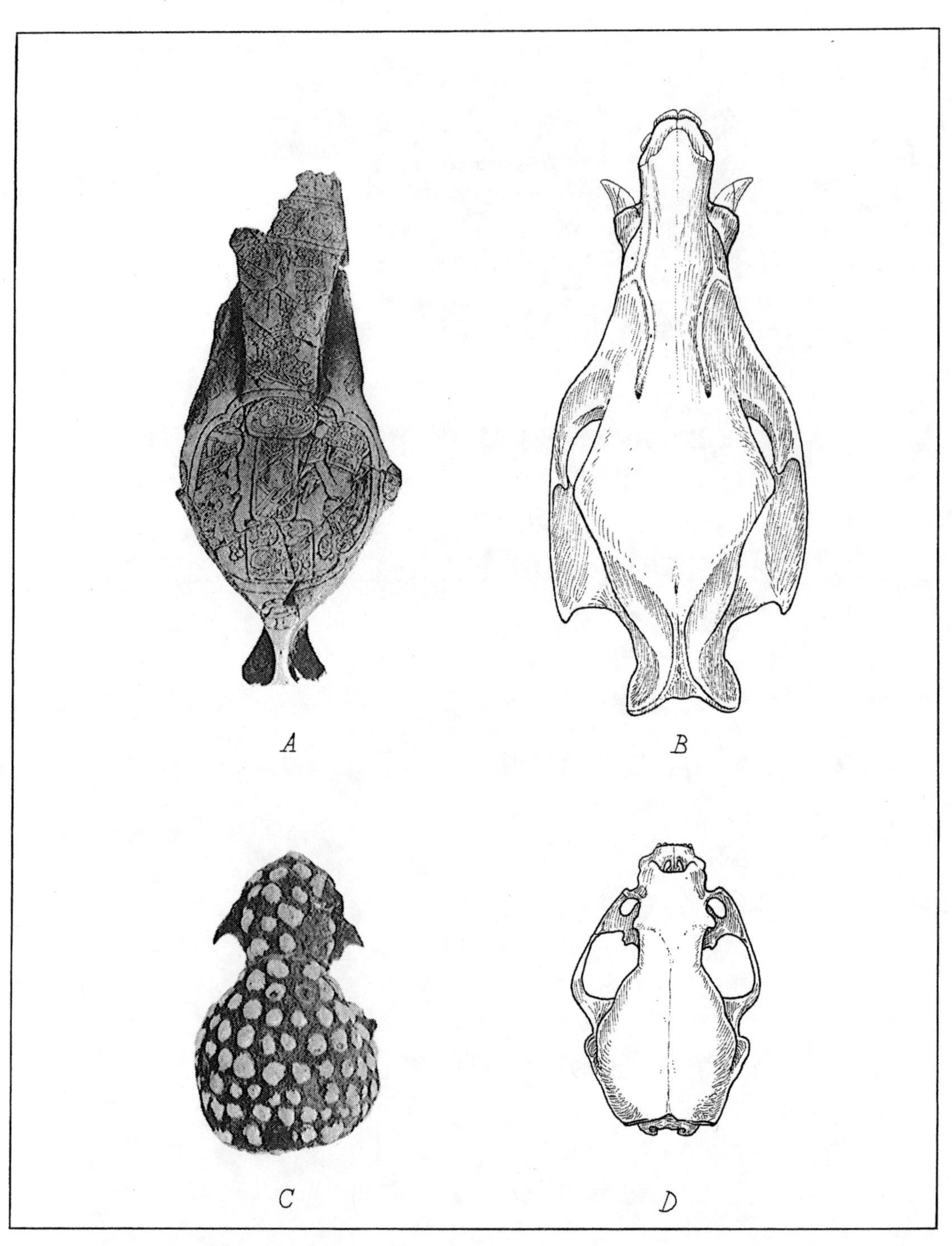

Examples of extreme modification of animal bones by carving or adornment. *A*, Engraved cranium and rostral portion of peccary skull from Copan, Honduras. Peabody Museum No. C/201. *B*, Dorsal view of unaltered peccary skull. *C*, Otter skull lime container, decorated with pitch and mollusk shells, from Huara Valley, Peru. Peabody Museum No. 46-77-30/6294. *D*, Dorsal view of unaltered otter skull.

PAPERS
OF THE
PEABODY MUSEUM OF ARCHÆOLOGY AND
ETHNOLOGY, HARVARD UNIVERSITY
VOL. 56, No. 1

MAMMAL REMAINS FROM ARCHÆOLOGICAL SITES

PART I

SOUTHEASTERN AND SOUTHWESTERN UNITED STATES

BY

STANLEY J. OLSEN

PUBLISHED BY THE PEABODY MUSEUM
CAMBRIDGE, MASSACHUSETTS, U.S.A.

First published 1964
Second printing 1973
Third printing 1980
Fourth printing 1985
Fifth printing 1990
Sixth printing 1996
Seventh printing 2004

ISBN 0-87365-162-6
Library of Congress Catalog Card Number 85-176642
Printed by the Office of the University Publisher
Harvard University
Cambridge, Massachusetts, U.S.A.

ACKNOWLEDGMENTS

I WISH to especially thank Miss Barbara Lawrence of the Museum of Comparative Zoology at Harvard University, for the loan of the greater portion of the necessary comparative material used to complete this study and also for the use of the laboratory facilities at the Harvard museum, where much of this work was undertaken.

The drawings were made by the writer from camera lucida projections. The finished skull drawings are the work of the late Mr. Andrew Janson, scientific illustrator for the Florida Geological Survey. The postcranial bones are accurately represented by photographic line drawings. These are the result of painstaking work by Mr. Gerrit Mulders of Tallahassee.

Dr. R. O. Vernon, Director of the Florida Geological Survey, was especially generous in allowing the necessary time to do that portion of the research not compensated for by the National Science Foundation funds.

The following departmental curators generously lent comparative material from the collections under their care: Drs. D. Hoffmeister of the Museum of Natural History at the University of Illinois; J. C. Moore of the Chicago Natural History Museum; R. Kellogg and D. H. Johnson of the U.S. National Museum; F. Blair of the Department of Zoology at the University of Texas; J. S. Findley of the Biology Department at the University of New Mexico; R. Van Gelder of the Mammal Department at the American Museum of Natural History and A. C. Zeigler of the Museum of Vertebrate Zoology at the University of California.

I am grateful to Dr. A. C. Spaulding, Program Director of the Anthropology Section of the National Science Foundation, for his interest in this problem and for his assistance in obtaining the necessary research funds to complete this study under Grant No. G–17902. Publication costs were greatly defrayed by support of a National Science Foundation grant.

CONTENTS

LIST OF ILLUSTRATIONS

INTRODUCTION

In recent years several committees, composed for the most part of anthropologists and archæologists, have met to discuss ways and means of interpreting nonartifactual materials that turn up in archæological sites (Heizer and Cook, 1960; Taylor, 1957). One particularly important phase of this problem that has been discussed at all of these meetings is the need for basic identification field manuals for materials that the archæologist does not feel qualified to interpret. Identification of bones, other than those of man, has long plagued the archæologist. A canvass of the heads of many of the anthropology departments of universities in this country has indicated that the present style of publication was needed and would be utilized by field and laboratory workers as an aid toward identifying the bones encountered during the course of excavating.

It would be naïve to suppose that any single manual could be designed for the archæologist so as to eliminate the entire workload from the museum mammalogist or vertebrate palæontologist. That is not the intended purpose of this effort. Rather a manual should expedite the rough sorting of osteological material by the archæologist or by competent student help. In this way the collections might be more readily accepted by the specialist who generally shies away from the time-consuming preliminary rough work that must be done before the actual close comparisons can be made to determine which animals are present at the site.

It is difficult to draw the line regarding specific animals to be included in a study, or to decide what species are to be eliminated. To include all controversial forms would not be feasible because of the numerous subspecies that are to be found in some categories, for example in the rodent genus *Geomys*. Many animals, particularly insectivores, chiropterids, cetaceans, pinnipeds, and sirenians, are not encountered in sufficient quantity in the average archæological dig to justify their being included in a general field manual. If the present publication proves to be of practical use to the field archæologist, representatives of these families can be covered in a future publication.

The skeletons of the bison and domestic cattle have been illustrated and compared in an earlier work (Olsen, 1960).

In many instances where there are single diagnostic features that key out specific animals it will be possible for the archæologist to arrive at exact identifications from this manual alone. For example, the inflected angle of the lower jaw is peculiar to the opossum *Didelphis marsupialis* (fig. 5), and the peglike teeth and extended rostrum of the skull are characteristic of the armadillo *Dasypus novemcinctus* (fig. 5).

Where the problem arises as to whether the forms in question are introduced domestic or native wild stock, only the skulls and dentitions have been illustrated and discussed in detail (for example the goats *Capra hirca* and *Oreamnos americanus*). The published results of a statistical study of the variations observed in the postcranial skeletons of each of these animals would be of monographic proportions and, in the majority of cases, the numerous skeletons required for a thorough evaluation of this sort are not available in the known study collections. Where positive key features are known, from the few published results of previous workers who have dealt with this problem, they are indicated on the drawings and the references are cited in the text.

In a work of this kind, no matter how many visual aspects of each bone are presented, it seems that the person making the comparisons from a publication needs an additional view other than those which are shown. It was beyond the limits of this project to depict all of the surfaces and details of each element. In some instances it was not even practical, for example in some rare forms where the fusing of the sacrum with the pelvis made ventral photography of the sacrum impossible.

Where elements were deemed too small to be identified by all but experts (some field mice with vertebrae of 3 mm. length) they were not illustrated.

There are numerous mammal classifications in print. Those by Hall and Kelson (1959), Miller and Kellogg (1955), and Simpson (1945) are perhaps the ones most generally referred to by taxonomists. All of these works differ from one another to some degree, and in some instances add to the taxonomic confusion that exists by standing on the rules of priority in order to change such well-known generic names as *Odocoileus* for the deer to *Dama*, or to change *Ursus* for the black bear to *Euarctos*. The classification used here is that which is contained in the single volume publication by Miller and Kellogg (1955). This is a relatively inexpensive book, readily available to all readers who have need for such a reference, and is the result of thorough and exhaustive research.

It must be strongly emphasized that when making comparisons, by referring to the plates, the size of the element should be kept in mind at all times. In some of the smaller mammals, such as the rodents, the drawings of the bones have been enlarged while those of the elk and moose have been reduced considerably. A convenient millimeter scale is included with each drawing in order that the actual size of each component may be visualized. The overall size of some elements is sufficient evidence to narrow down the category to which they belong (length of moose skull among the artiodactyls). The dental formula should be particularly noted as the presence or absence of a premolar may be of taxonomic importance (bobcat skull compared with that of the puma).

I concur with the generally held opinion that an author is solely responsible for any shortcomings or omissions of text that may appear under his name. However, I reserve the right to one exception. During the four years, in which the notes for the present volume were being compiled, I canvassed many experts in mammalian osteology for additional information that could be included as keys to aid in the identification of fragmentary animal bones. I must report that only a small percentage of the sometimes promised and always needed information was forthcoming. I cannot stress too strongly that one purpose of this publication is to inspire other workers to put into print information of the sort contained here so that it can be utilized by the archæologist who otherwise *would not know of this hidden knowledge.*

MAMMALS REPRESENTED IN THIS STUDY

MARSUPIALIA
- Didelphiidae
  - *Didelphis marsupialis*, Opossum

EDENTATA
- Dasypodidae
  - *Dasypus novemcinctus*, Armadillo

LAGOMORPHA
- Ochotonidae
  - *Ochotona princeps*, Pika
- Leporidae
  - *Lepus californicus*, Black-tailed Jackrabbit
  - *Sylvilagus audubonii*, Desert Cottontail

RODENTIA
- Sciuridae
  - *Marmota flaviventris*, Yellow-bellied Marmot
  - *Cynomys gunnisoni*, White-tailed Prairie Dog
  - *Citellus variegatus*, Rock Squirrel
  - *Citellus spilosoma*, Spotted Ground Squirrel
  - *Eutamias dorsalis*, Cliff Chipmunk
  - *Sciurus griseus*, Western Gray Squirrel
  - *Sciurus niger*, Eastern Fox Squirrel
  - *Sciurus aberti*, Tassel-eared Squirrel
- Geomyidae
  - *Geomys bursarius*, Plains Pocket Gopher
- Heteromyidae
  - *Perognathus apache*, Apache Pocket Mouse
  - *Dipodomys ordii*, Ord Kangaroo Rat
- Castoridae
  - *Castor canadensis*, Beaver
- Cricetidae
  - *Oryzomys palustris*, Rice Rat
  - *Reithrodontomys megalotis*, Western Harvest Mouse
  - *Peromyscus crinitus*, Canyon Mouse
  - *Peromyscus maniculatus*, Deer Mouse
  - *Onychomys leucogaster*, Northern Grasshopper Mouse
  - *Sigmodon hispidus*, Hispid Cotton Rat
  - *Neotoma albigula*, White-throated Woodrat
  - *Microtus montanus*, Mountain Vole
  - *Ondatra zibethicus*, Muskrat
- Zapodidae
  - *Zapus princeps*, Western Jumping Mouse
- Erethizontidae
  - *Erethizon dorsatum*, Porcupine

CARNIVORA
- Canidae
  - *Canis familiaris*, Domestic Dog
  - *Canis latrans*, Coyote
  - *Canis lupus*, Wolf
  - *Vulpes fulva*, Red Fox
  - *Vulpes macrotis*, Kit Fox
  - *Urocyon cinereoargenteus*, Gray Fox
- Ursidae
  - *Euarctos americanus*, Black Bear
  - *Ursus horribilis*, Grizzly Bear
- Procyonidae
  - *Bassariscus astutus*, Cacomistle
  - *Procyon lotor*, Raccoon
  - *Nasua narica*, Coati
- Mustelidae
  - *Martes americana*, Marten
  - *Mustela frenata*, Long-tailed Weasel
  - *Mustela vison*, Mink
  - *Mustela nigripes*, Black-footed Ferret
  - *Taxidea taxus*, Badger
  - *Mephitis mephitis*, Striped Skunk
  - *Conepatus leuconotus*, Hog-nosed Skunk
  - *Lutra canadensis*, River Otter
- Felidae
  - *Felis onca*, Jaguar
  - *Felis pardalis*, Ocelot
  - *Felis concolor*, Puma
  - *Felis yagouaroundi*, Yagouaroundi
  - *Lynx rufus*, Bobcat

PERISSODACTYLA
- Equidae
  - *Equus caballus*, Domestic Horse

ARTIODACTYLA
- Tayassuidae
  - *Sus scrofa*, Domestic Pig
  - *Pecari tajacu*, Peccary
- Cervidae
  - *Cervus canadensis*, Elk (Wapiti)
  - *Odocoileus hemionus*, Mule Deer
  - *Odocoileus virginianus*, *White-tailed Deer*
  - *Alces americana*, Moose
- Antilocapridae
  - *Antilocapra americana*, Prong-horned Antelope
- Bovidae
  - *Capra hirca*, Domestic Goat
  - *Oreamnos americanus*, Mountain Goat
  - *Ovis aries*, Domestic Sheep
  - *Ovis canadensis*, Mountain Sheep

MAMMAL REMAINS FROM ARCHÆOLOGICAL SITES

PART I

SOUTHEASTERN AND SOUTHWESTERN UNITED STATES

THE MAMMAL SKULL

Most differentiating characters, separating mammals on a specific level, have to do with size, color, markings or other prominent external features generally seen in the living animal. Osteologically speaking, specific characters are mostly limited to the skull and dentition.

There are various diagnostic characters which are peculiar to the skulls of mammals. There is a double occipital condyle for the articulation with the first cervical vertebra. There is a secondary hard palate of bone separating the nasal passage to the mouth. The external narial passage is a single opening in the front of the skull. The zygomatic arch stands out from the sides of the skull. The mammals have a differentiated dentition consisting of incisors, canines and cheek teeth that have a crown consisting of several cusps. There are rare exceptions to this rule as in the armadillo. There are two definite tooth generations and only marginal teeth are present in the jaws. The ramus of the mandible is a single bone. The jaw articulates directly with the squamosal.

Certainly the teeth of an animal are the most important features of the skull used in the classification of these forms. This is only natural, since the teeth are adapted to the animal's food habits and thus show a great variety of form and structure among the many mammals that are found in archæological sites (fig. 1).

The amount of wear and the shape of the individual tooth are also indicators of an animal's age (Olsen, 1961b), an important factor in determining what age groups were utilized for food or utilitarian purposes by primitive peoples (fig. 2).

The skulls are discussed in the order in which they are classified by mammalogists. Illustrations of skulls of similar shape and form are kept together as nearly as possible to make the task of comparison easier for the non-mammalogist.

So numerous and varied are rodent teeth that no two authors agree on the nomenclature of the cusps. Some workers feel that the more complex patterns defy analysis and apply such terms as, "metaconid-metastylid chaos" to these areas (McKenna, 1962). The designations used here are those that are most generally accepted by rodent specialists.

FIGURE 1

The major tooth forms and cusp nomenclature. So numerous and varied are the rodent teeth that no two authors agree on the nomenclature of the cusps. The designations used here are those that are most generally accepted by workers in this field.

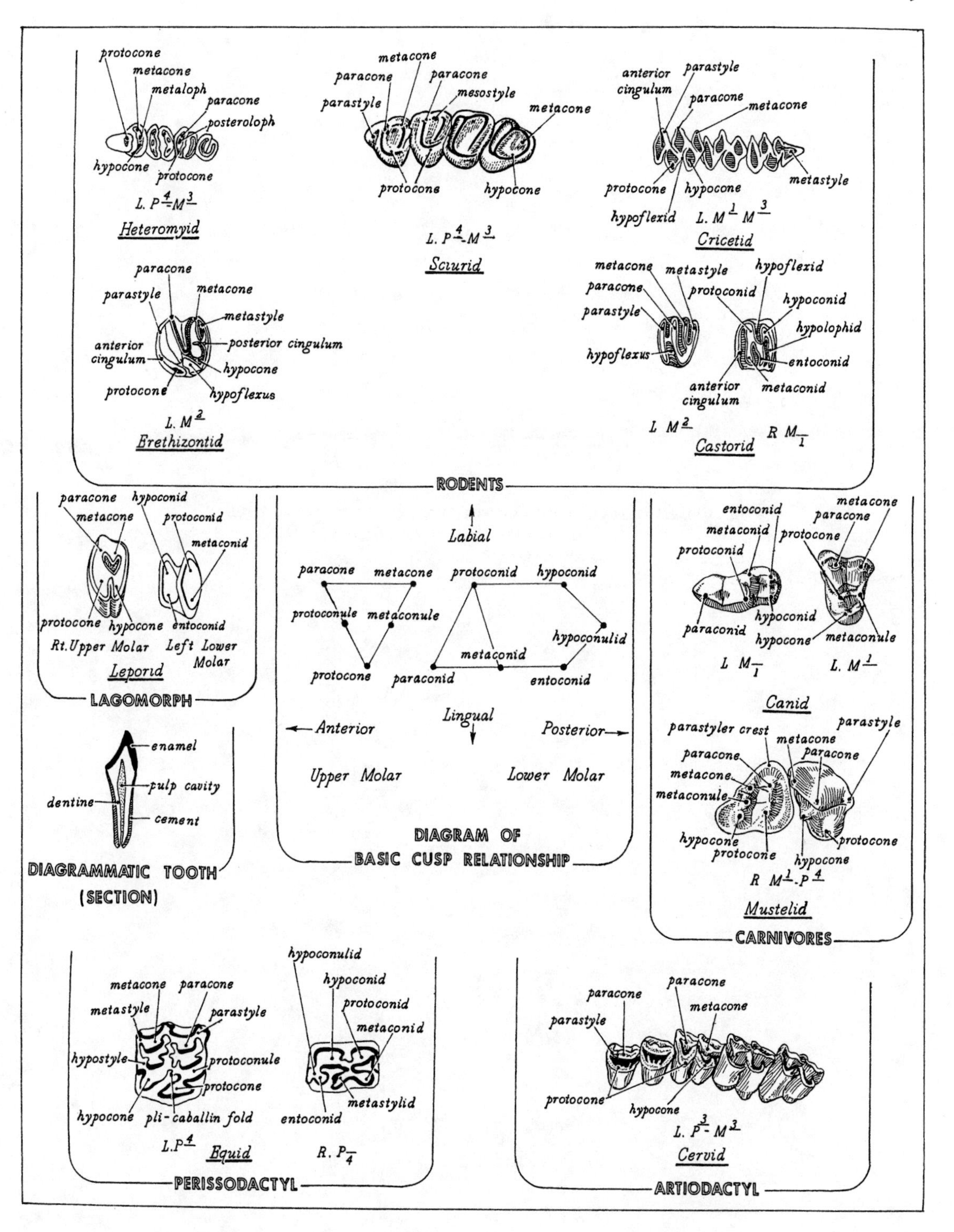
protocone
metacone
metaloph
paracone
posteroloph
hypocone
protocone
L. P4-M3
Heteromyid
metacone
paracone
paracone
parastyle
mesostyle
metacone
protocone
hypocone
L. P4-M3
Sciurid
anterior cingulum
parastyle
paracone
metacone
protocone
hypocone
metastyle
hypoflexid
L. M1 M3
Cricetid
paracone
parastyle
metacone
metastyle
anterior cingulum
posterior cingulum
hypocone
protocone
hypoflexus
L. M2
Erethizontid
metacone
metastyle
hypoflexid
paracone
protoconid
hypoconid
parastyle
hypolophid
hypoflexus
entoconid
anterior cingulum
metaconid
L M2
R M1
Castorid
RODENTS
paracone
hypoconid
metacone
protoconid
metaconid
protocone
hypocone
entoconid
Rt. Upper Molar
Left Lower Molar
Leporid
LAGOMORPH
Labial
paracone
metacone
protoconid
hypoconid
protoconule
metaconule
hypoconulid
metaconid
protocone
paraconid
entoconid
Lingual
Anterior
Posterior
Upper Molar
Lower Molar
DIAGRAM OF BASIC CUSP RELATIONSHIP
enamel
pulp cavity
dentine
cement
DIAGRAMMATIC TOOTH (SECTION)
entoconid
metacone
paracone
metaconid
protocone
protoconid
hypoconid
paraconid
hypocone
metaconule
L M1
L. M1
Canid
parastyler crest
parastyle
metacone
paracone
paracone
metacone
metaconule
hypocone
protocone
protocone
hypocone
R M1-P4
Mustelid
CARNIVORES
hypoconulid
hypoconid
metacone
paracone
protoconid
metastyle
parastyle
metaconid
hypostyle
protoconule
protocone
metastylid
hypocone
pli-caballin fold
entoconid
L.P4
Equid
R. P4
PERISSODACTYL
paracone
paracone
parastyle
metacone
protocone
hypocone
L. P3-M3
Cervid
ARTIODACTYL

FIGURE 2

Various stages of incisor wear as seen in the Elk, *Cervus canadensis*.

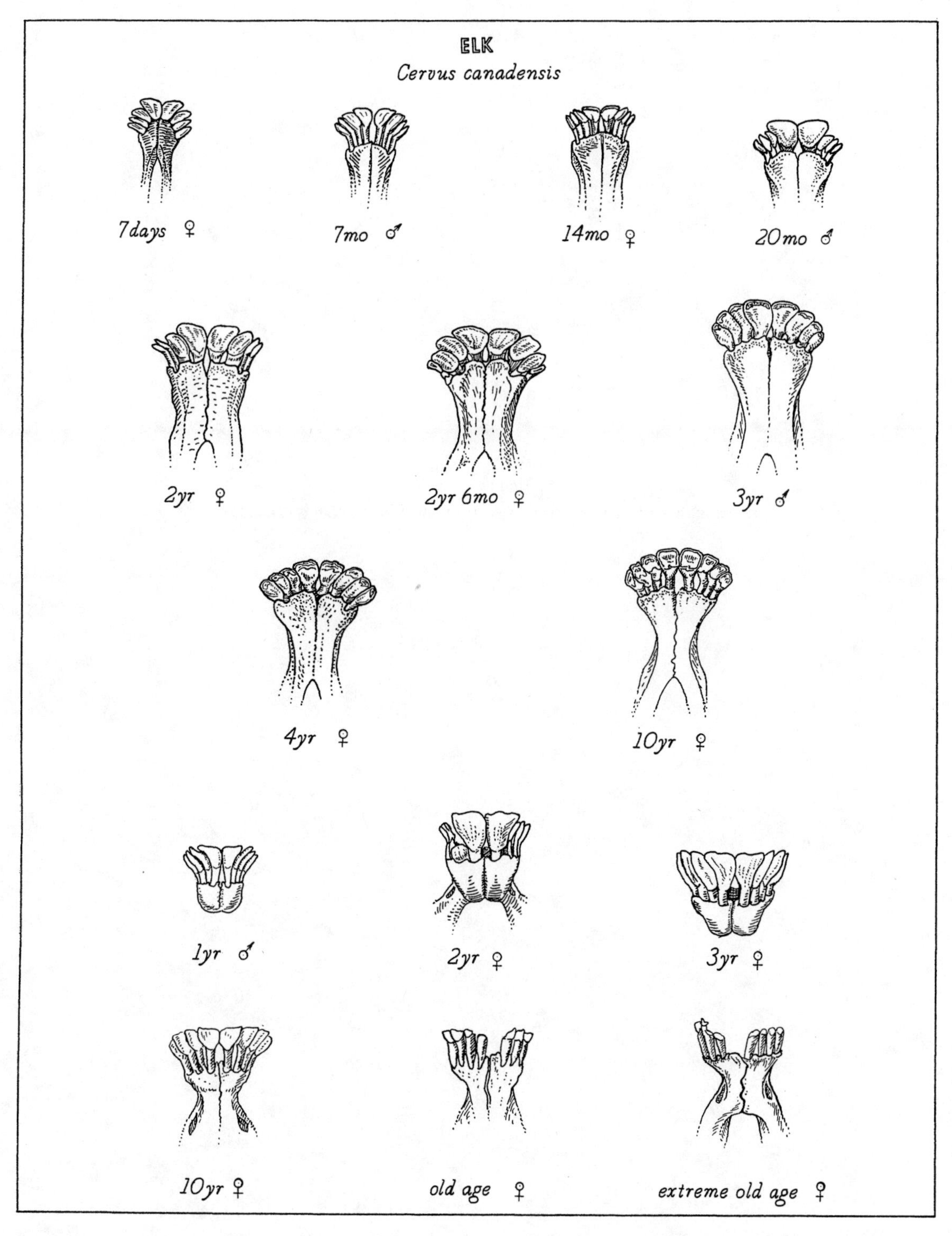
ELK
Cervus canadensis
7days ♀
7mo ♂
14mo ♀
20mo ♂
2yr ♀
2yr 6mo ♀
3yr ♂
4yr ♀
10yr ♀
1yr ♂
2yr ♀
3yr ♀
10yr ♀
old age ♀
extreme old age ♀

FIGURE 3

Various stages of incisor wear as seen in the Horse, *Equus caballus*.

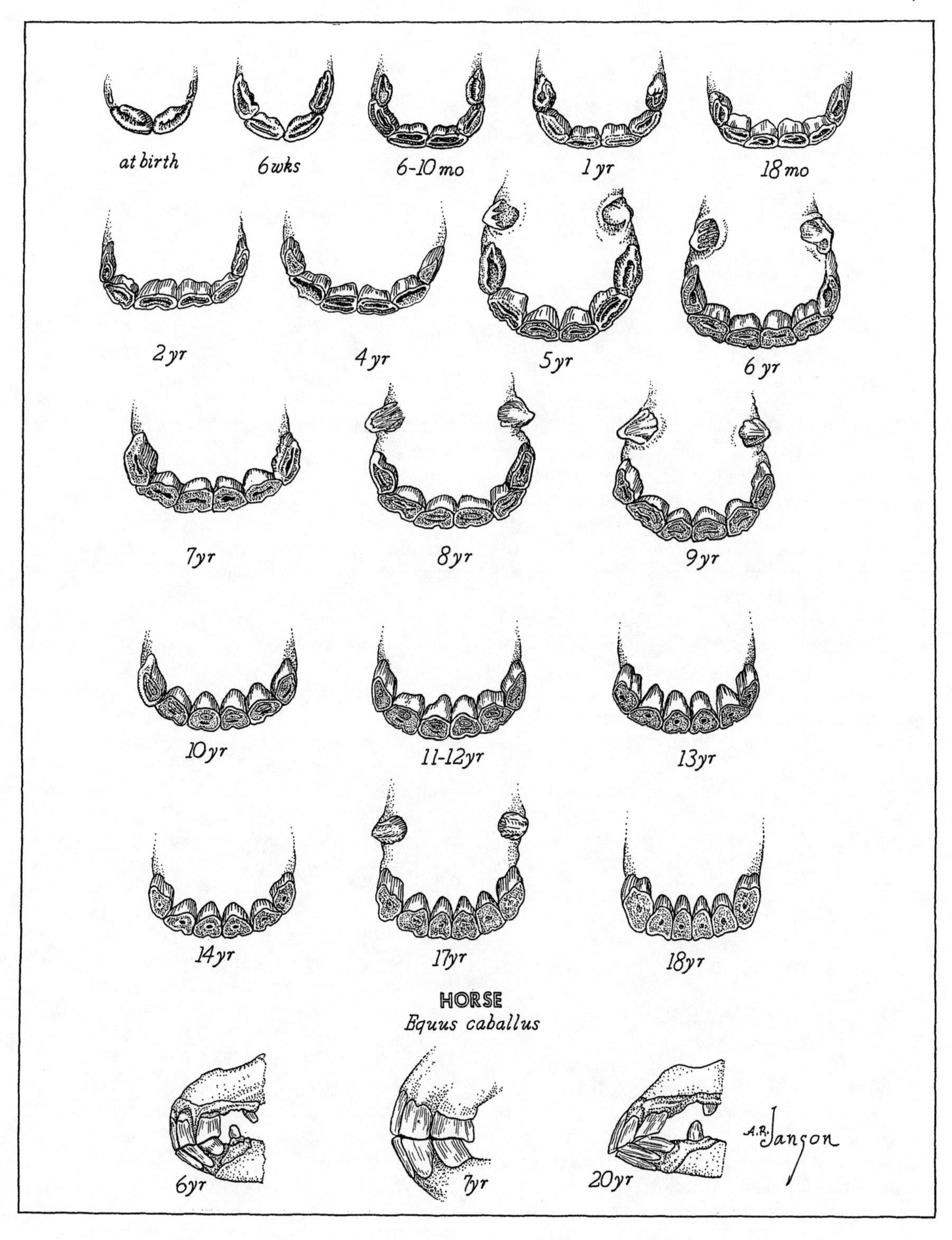

HORSE
Equus caballus

GENERAL FEATURES OF THE SKULL IN DIFFERENT GROUPS OF MAMMALS

(Specific differentiating characters are indicated or noted on the plates depicting the individual skulls and dentitions.)

Marsupialia

Didelphiidae: Only one species is considered under this family, the opossum *Didelphis marsupialis*. The skull has a proportionately smaller braincase and a high saggital crest in old individuals. The jugal contributes to the glenoid fossa and the tympanic is rarely fused with other bones. The auditory bulla is formed entirely from the alisphenoid. The middle ear is not enclosed by the tympanic to form the auditory bulla but is partially enclosed by a cup-like process of the alisphenoid. The palatines are perforated in several places to form palatine openings. The palatine has a square posterior margin. The dental battery consists of 50 teeth (average of 40 in other mammals). There are five incisors above and four in the lower jaw. There are four lower molars. The upper and lower molars are tritubercular with well-developed trigonids. The lower jaw has a distinguishing inflected angle.

There is considerable size variation in the adult animals, much more so than is observed in other similar-sized small mammals.

Edentata

Dasypodidae: As with the opossum, only one species of armadillo *Dasypus novemcinctus* is considered in this family. The nasal region is elongated. The pterygoids meet below the posterior nares and the anterior opening of the infraorbital canal is nearer to the posterior ends of the pterygoids than to the anterior ends of the nasals. The teeth are rootless, enamelless pegs with no differentiation between the premolars and molars. The tooth row terminates anteriorly to the root of the zygoma. There are no teeth at the symphysis of the lower jaw or in the region of the premaxilla.

Lagomorpha

A peculiar lattice-like fenestration of the side of the skull is a noticeable feature of the rabbits. The palate is short, not extending to the end of the tooth row.

Ochotonidae: The pika, *Ochotona princeps*, has no supraorbital process on the frontals. The nasals are widest anteriorly. The maxilla has a single large fenestra. The skull is flattened, with a wide interorbital region. The maxillary oriface is roundly triangular. The palatal foramina are separate from the anterior palatine foramina. The malar is prolonged almost to the auditory opening. The angle of the lower jaw terminates well back of $M/_3$. The teeth have persistent pulp and continuous growth. There is one less cheek tooth above. The second upper maxillary tooth ($P_4/$) is unlike the third in form. The last lower molar ($M/_3$) has more than one re-entrant angle. The incisors are long, broad, with a deep groove on the outer side (dividing the tooth into unequal parts), terminating in two sharp points. There is a second small pair of incisors set directly behind the first. The incisors and molariform teeth are separated by a wide diastema. There are three upper incisors in newly born pikas.

Leporidae: In contrast with the rodents, the two upper rows of teeth are farther apart than the lower ones, giving the teeth more of a chopping motion than that of grinding. The grinding surfaces of the cheek teeth have an "H" shaped pattern. The optic foramina are united to form a single opening. The coronoid process is slightly differentiated from the ascending portion of the mandible.

The black-tailed jackrabbit, *Lepus californicus* has the paroccipital process terminating outward from a noticeable swelling in the parietal area. The nasals are wider anteriorly as in the pika. The interparietal is fused to the surrounding bones. The posterior wing of the supraorbital process is free of the braincase. The bullae are proportionately smaller than in the desert cottontail. There is a bony network on the side of the skull, anterior to the orbit. The backward projection of the zygomatic arch is common to rabbits but does not extend as far in the pika. The palate forms a narrow bridge. There are many perforations on the back of the skull which are not present in the cottontail rabbit. The supraorbital is subtriangular in outline. There is a wide diastema between the incisors and the cheek teeth. Two pairs of incisors are in tandem but with square cutting edges instead of terminating in a double point as in the pika. The groove on the outer face of the outer incisors separates the surface into unequal parts. In lagomorphs the enamel extends to the posterior side of the incisors. In the rodents the enamel is confined to the anterior face.

The desert cottontail, *Sylvilagus audubonii*, has paroccipital processes that lie close against the surface of the bullae. The nasals are not wide anteriorally as in the pika. The tympanic bullae are large. The interparietals are plainly visible. The posterior wing of the supraorbital process touches or is fused to the braincase. As in the jackrabbit, a bony network is present on the sides of the skull, anterior to the orbits. The backward projection of the zygomatic arch is present as in the jackrabbit. The palate forms a narrow bridge. There is an absence of perforations at the base of the skull (compared to the jackrabbit). Auditory meatus measures more than 37 percent of the alveolar length of the upper tooth row. The supraorbital is straplike. There is a wide diastema between the incisors and the cheek teeth. As in the jackrabbit, there are two pairs of upper incisors in tandem, but these teeth are square on the cutting edge instead of terminating in a double point as in the pika. Grooves separate the anterior face of the outer incisors into nearly equal halves.

Rodentia

In rodents the orbits open posteriorly into the temporal fossa. There is often a large paroccipital process, infraorbital foramen and anterior palatal foramina. The jugal bone is well defined. The lachrymal extends considerably on the cheek, anterior to the orbit. A small coronoid may be present or absent on the mandible. The angle of the lower jaw is well developed. The mandibular condyle is rounded or oval-shaped. There is often a prominent masseter crest on the horizontal ramus. The lachrymal foramen is always well within the margin of the orbit. The premaxilla reaches to the frontal. The rodents have a single pair of incisors opposed by a principal pair above. These teeth never form roots and have bases which curve far back inside the bones of the upper and lower jaws. Their continual growth counterbalances the wear to which they are subjected in gnawing. Loss of other incisors, canines and a number of premolars leave a long diastema between the gnawing teeth and the grinding series. There are never more than two upper premolars and a single lower one. In many rat-like forms only the molars remain in the cheek region. The cheek teeth often become high crowned and in many cases they fail to close their roots and like the incisors continue to grow throughout life. The upper rows of grinders are tilted so that they meet the lower teeth in a plane that faces somewhat outward as well as downward.

Sciuridae: The channel for the masseter lies anterior to the zygomatic arch and extends far up on the side of the rostrum. Distinct postorbital processes are present. The infraorbital opening is small. The postorbital process is at right angles to the long axis of the skull. The teeth are cuspate and rooted. A postorbital process is present on the frontal bone. The sciurids have a broad zygomatic plate and a well-developed zygomatic arch. The skull of the yellow-bellied marmot, *Marmota flaviventris*, has a narrow interorbital region. The postorbital processes project back of a line drawn across their bases and at right angles to the long axis of the skull. The nasals are no broader posteriorly than the premaxillae. The posterior border of the palate is bevelled at an obtuse angle. The incisive foramen is constricted posteriorly or of equal width throughout. The maxillary tooth rows are slightly divergent anteriorly. The skull of the marmot has a nearly flat top which forms a straight line when viewed laterally. The postorbital processes of the frontal are triangular in shape.

The skull of the white-tailed prairie dog, *Cynomys gunnisoni*, is broad and robust with a well-developed occipital crest. The sagittal crest is moderately developed anteriorly, but more developed posteriorly. The squamosal root of the zygomatic arch is wide and spreading. The antorbital foramen is subtriangular with a prominent tubercle. The tooth rows are strongly convergent posteriorly. The individual cheek teeth are large and expanded laterally. The first premolar is large and nearly equal to the second. $M^3/$ has an additional transverse ridge. The white-tailed prairie dog differs from other species by having a more broadly spreading maxillary arch and mastoids smaller and more obliquely placed. The bullae are smaller and there is a stout, decurved postorbital process. The jugal is weak, thin and flat, the outer surface at the angle of the ascending branch is only slightly thickened and has a rounded margin. The teeth of *C. gunnisoni* are smaller and less expanded laterally when compared with other species of this genus.

The skull of the rock squirrel, *Citellus variegatus*, is less flattened than in the larger *Marmota*. The postorbital process of the frontal projects backward to a greater degree. The skull differs from *Sciurus* in having a shallower braincase, zygoma that are more twisted anteriorly and the anterior portion of each zygoma twisted toward a horizontal plane. The upper incisors are more compressed laterally and more nearly flat on the anterior faces. The rock squirrel skull differs from *Eutamias* in having the anterior border of the zygomatic notch in the maxillary opposite $M^1/$ rather than $P^4/$. The masseteric tubercle is of medium size, situated ventral and slightly lateral to the infraorbital canal. In *Citellus* there is an infraorbital canal as opposed to a foramen in *Eutamias*. A definite bend is present on the dorsal profile of the skull at the junction of the rostrum and the cranium when viewed laterally. The zygomatic arches are expanded posteriorly but not as oppressed as in *Eutamias*, and not as wide as in *Marmota*. The upper tooth rows are nearly parallel. *C. variegatus* differs from other species in its larger size, and in having narrower oval openings to the infraorbital canal. $P^3/$ is about one-sixth the size of $P^4/$. The skull has a flattened dorsal outline and a relatively broad, shallow braincase. The rostrum is relatively broad, tapering gradually. The postorbital processes are stout and decurved. The skull of the spotted ground squirrel, *Citellus spilosoma* is closest to *C. tridecemlineatus* of this genus but is relatively broader, especially in the rostral and interorbital region. The auditory bullae are much larger. The width at the postorbital constriction is slightly more than the interorbital width. The fossae that are anterolateral to the incisive foramina are deep. Upper cheek teeth are low-crowned with $M^1/$ and $M^2/$ subquadrate in occlusal outline. $M^3/$ is slightly larger than $M^2/$.

The skull of the cliff chipmunk, *Eutamias dorsalis*, is slightly built and narrow with light and weakly constructed postorbital processes. The lacrimal is not elongated, the infraorbital foramen lacks a canal and is relatively larger than most sciurids. $P^3/$ is present. Foramen are found instead of an infraorbital canal as in *Citellus*. The anterior border of the zygomatic notch in the maxillary is opposite $P^4/$ instead of opposite $M^1/$ as in *Citellus*. The transverse ridges diverge labially, whereas they are more nearly parallel in the tree squirrels. $P^3/$ is a mere spike. The palate terminates on the plane of the last molars or a little posterior to it. The zygomatic plate of the maxillary is usually opposite the middle or posterior part of the last premolar. The bullae are relatively large. There are slender, backward directed postorbital processes. The interorbital foramen is oval in shape. *E. dorsalis* differs from other species of *Eutamias* by having the top of the braincase wide and flattened but less than in *E. panamintus*. The incisive foramina diverge posteriorly rather than being parallel. The rostrum is relatively broad.

The skull of the western gray squirrel, *Sciurus griseus*, has the following generic characters. The squamosal is low and extends up the cranium no more than about halfway from the posterior edge of the suborbit to the base of the postorbital process of the frontal. The postorbital process is well developed. The upper anterior premolar is small when present. The anterior border of the orbit is ventrally opposite the first molar. The jugal has an angular process on its upper surface. The palate is nearly square posteriorly and terminates immediately behind the tooth rows. The infraorbital always forms a canal. The masseteric tubercle is weakly present. $M^1/$ and $M^2/$ are present with four transverse crests. The comparative total length of the skull should be noted, especially as a distinguishing character for the fox and tassel-eared squirrels.

S. griseus differs from other species of the genus in having weak jugals with relatively little twist from the vertical plane. The skull is broad, especially across the parietals. The nasals terminate posteriorly, subequally with the posterior tongues of the premaxillae. The molars are massive. The second premolar is peglike.

The skull of the eastern fox squirrel, *Sciurus niger*, lacks $P^3/$ and has only one upper premolar. There are four upper cheek teeth (as compared to five in other species of *Sciurus*). The frontals are slightly elevated posteriorly. The interorbital notch is distinct. $P^4/$ is subtriangular as compared to quadrangular in cross section. The infraorbital foramen is small.

The skull of the tassel-eared squirrel, *Sciurus aberti*, possesses five upper cheek teeth (as does *S. griseus*). The skull is short and broad with a flattened frontal area and a wide and depressed braincase. The rostrum is narrow and laterally compressed. The nasals are long and equal to the interorbital breadth. There are two upper premolars and one small fourth lower premolar.

Geomyidae: The skull is massive and angular and not greatly enlarged in the auditory region. The infraorbital canal is long and narrow and opens on the lateral side of the rostrum, anterior to the zygomatic plate. The squamosals possess strong ridges which unite as a sagittal crest. The zygomatic arches are strong. The cheek teeth are ever-growing. The molars have enamel plates on either the anterior or posterior surfaces.

The skull of the plains pocket gopher, *Geomys bursarius*, is heavy and triangular, essentially flat from above the occiput to near the anterior ends of the nasals. The zygoma is robust and widespread, the anterior border abruptly angular to the lateral arches and to the rostrum. The frontal is without a postorbital process. The rostrum is prominent, long and heavy, with sides that are nearly parallel. The nasals are long and relatively narrow and in the upper view are only slightly broader anteriorly. The bullae are small, angular, and not inflated with a length about twice that of the width. The external auditory meatus is elongated and tubular. $P^4/$ has only three enamel plates and is decidedly larger than $P/_4$. The first and second upper molars each have an anterior and a posterior plate. The lower molars have a single posterior plate. The upper incisors have two longitudinal grooves, a major medial groove and a smaller inner groove.

G. bursarius differs from other species in that the nasals are not hourglass-shaped and there is little or no constriction near the middle. The rostrum is wider than the basioccipital is long.

Heteromyidae: The heteromyid skull is thin-walled and has a highly inflated tympanic region. There is an infraorbital canal as in the geomyids. The cheek teeth are rooted. The incisors are thin and compressed.

The skull of the apache pocket mouse, *Perognathus apache*, has an ethmoid foramen in the frontal. The ventral surface of the tympanic bulla is below the level of the grinding surface of the upper cheek teeth and reaches the upper side of the skull. The zygoma have roots free from the bullae. The auditory bullae are more or less triangular in outline, anteriorly opposed to pterygoids. The mastoids are expanded and are visible in dorsal aspect. The jugal is light and threadlike. The rostrum is slender and pointed. The infraorbital foramen is a small opening in the maxilla. The lophs of the upper molars unite progressively from lingual to buccal margins, while those of the lower premolars unite at the center of the tooth, giving an "X" pattern. The lophs of the lower molars unite primitively at the buccal margin; progressively at the center of the tooth forming an "H" pattern. The cheek teeth are rooted. The incisors are grooved.

P. apache differs from other species by having a short upturned angular process of the mandible. The skull is short and broad and the nasals are relatively long. The interparietal is small and nearly as long as wide. The lower premolar is smaller than the last molar. The mastoids are large and the bullae are opposed anteriorly.

The skull of the ord kangaroo rat, *Dipodomys ordii*, is triangular with a small interparietal and a slender jugal. The ventral surface of the bullae barely reaches the level of the grinding surfaces of the cheek teeth and sometimes is below that level. There is no ethmoid foramen in the frontal. The zygomatic root of the maxilla is expanded anteroposteriorly. The center of the palate, between the premolars, is ridged. The pterygoid fossae are double. The bullae and other bones associated with the ear are greatly inflated. The bullae are exceeded in relative size only by members of the related genus *Microdipodops*. The nasals are long and overhanging. The bullae are hollow instead of cancellous as in the pocket mice. The cheek teeth have the "H" pattern present with enamel limited to the anterior and posterior plates. $M^3/_3$ are small. The upper incisor teeth have longitudinal grooves. There are four cheek teeth.

D. ordii differs from other species in that the lower incisors are awl-shaped instead of chisel-shaped. The rostrum is relatively short and the interparietals are wide. The zygomatic process of the maxillary is also wide.

Castoridae: The skull is massive with strong zygomatic arches and possesses a broad, deep rostrum. The upper molars are subequal, and are rooted and hypsodont. The incisors are strongly developed.

The skull of the beaver, *Castor canadensis*, has a massive, broad, deep rostrum and a narrow braincase. The basioccipital region has a conspicuous pit-like depression. The zygomatic arches are strong. The skull is without postorbital processes. The infraorbital canal is inconspicuous, opening on one side of the rostrum anterior to the zygomatic plate. The anterior opening of the infraorbital canal is much smaller than the incisive foramen. There is a decided angle at the symphysis of the mandible when viewed laterally. The mastoid process projects forward. The cheek teeth are rootless and exhibit alternating layers of enamel and dentine, with re-entering enamel folds, two on the outer and one on the inner side. The front surfaces of the incisors are orange-red in color, the lower pair longer with chisel-like edges. The cheek teeth are not ever-growing.

Cricetidae: In the cricetid skull the squamosals are greatly expanded, with corresponding reduction of the parietals and interparietals. Interparietal constriction of the frontals attains its greatest development. There is a short and slender jugal, usually as a splint between the zygomatic process of the maxilla and the squamosal. The lower root of the zygomatic process is flattened with a perpendicular plate. The infraorbital vacuity is tall and wide above and restricted below. The infraorbital canal has a rounded upper portion which transmits part of the masseter muscle and has a restricted lower part which transmits the nerve. There is no postorbital process. The cheek teeth are in parallel rows, laminate or prismic.

The skull of the rice rat, *Oryzomys palustris*, is slightly constricted and not markedly sculptured. The zygomatic arches are depressed to near the level of the tooth rows. The palate extends posteriorly beyond the tooth rows. Palatal pits are present. The bullae are moderately inflated. The anterior border of the lachrymal articulates about equally with both the maxilla and the frontal. There are usually prominent temporal ridges. Molars have well developed cusps, the main cusps arranged in two longitudinal, parallel rows. The upper incisors have a decided backward curve. The upper molars are three-rooted and the lower molars are double-rooted.

The skull of *O. palustris* differs from other species within the genus by having a larger skull with a high braincase and a short rostrum. The frontal region is broad, with trenchant lateral margins that are slightly upturned and project as supratemporal ridges. The temporal ridges are well developed. The interparietal is small and subtriangular. There are elongated palatal slits, approximately as large as the palatal bridge. Large sphenopalatine vacuities are present. A prominent sphenopalatine foramen is present behind the last upper molar. The palate extends beyond the last molar. There are two rows of tubercles on the molars. Three upper cheek teeth are present.

The skull of the western harvest mouse,

Reithrodontomys megalotis, is comparatively smooth and little sculptured with slender zygomae. The outer wall of the infraorbital foramen is a broad, thin lamina. The anterior palatine foramina are long and are separated by a thin septum, terminating approximately at the level of the anterior end of the tooth row. The posterior border of the palate is usually truncate, but often with a small median spine. The bullae are moderately inflated, arranged obliquely in relation to the long axis of the skull. The portion of the skull that is anterior to the least interorbital constriction is approximately equal in length to the portion that is posterior to the constriction. The braincase is moderately inflated, extending slightly beyond the limits of the zygomatic arches, when viewed dorsally. The anterorbital foramen is situated in the zygomatic portion of the maxillary, circular above, contracting to a slit below. The palate is square posteriorly. The coronoid process of the mandible is low. The upper incisors have a deep median groove. There are three rooted upper cheek teeth with cusps arranged in two parallel longitudinal rows.

R. Megalotis differs from other species by having a broad zygomatic plate and broad pterygoid fossae. The anterior zygomatic plate does not project anteriorly enough to be visible when viewed dorsally. The posterior border of the palate is truncate. The braincase is broad. There is a narrow rostrum. The dentine of the last upper molar is a continuous pattern and not separated into islands. The dentine of the last lower molar is arranged in a "C" pattern.

The skull of the canyon mouse, *Peromyscus crinitus*, is relatively little sculptured and thin-walled, with the interorbital constriction always apparent. The supraorbital ridges are poorly developed and sometimes beaded, sometimes trenchant. The rostrum is long and slender. Interparietals are well developed and the zygoma are delicate and depressed to the level of the palate. The zygomatic plate is narrow, approximately straight anteriorly. The anterior palatine foramina are long, slit-like and separated by a thin bony septum, with no palatal pits. The bullae are situated obliquely to the long axis of the skull. The nasals usually project forward over the incisors. The coronoid process of the mandible is strongly reduced. The upper incisors lack the pronounced grooves. There are three cheek teeth in each jaw, the anterior one being the largest, the posterior one being the smallest. The cusps form a distinct double row of tubercles. The cheek teeth are rooted.

The skull of *P. crinitus* differs from other members of the genus by having a short tooth row and the width across the anterior part of the zygomatic arch is less than the greatest width of the braincase. The premaxillae do not extend markedly posterior to the nasals. The zygomata are compressed anteriorly.

The skull of the deer mouse, *Peromyscus maniculatus*, differs from the others by being smooth and delicately built with a somewhat arched and well-inflated braincase. The infraorbital foramen is not visible from the side view, being covered by the zygomatic plate. The bullae are moderate to small. The rostrum is slender and tapered. There are paired posterior palatine foramina located midway between the anterior palatine foramina and the posterior edge of the palate. The anterior lateral border of the infraorbital plate is bowed forward. There is an absence of the ridge above the orbit. The palate ends opposite the last molars.

The skull of the northern grasshopper mouse, *Onychomys leucogaster*, has wedge-shaped nasals extending well beyond the premaxillary tongues. The interorbital constriction is narrow. There is a narrow zygomatic plate which is straight anteriorly. There are no supraorbital ridges. The bullae are small. The palate ends slightly behind the third molar. The coronoid process of the mandible is high (higher than in *Peromyscus*). The upper and lower molars have a double row of cusps.

Specific characters for *O. lencogaster* are: $M^3/$ is subcircular in cross section and $M^1/$ occupies less than one-half the length of the tooth row. The skull is relatively narrow interorbitally.

The skull of the hispid cotton rat, *Sigmodon hispidus*, has a well-developed rostrum and pronounced supraorbital ridges which extend posteriorly over the temporal region onto the parietals. The interparietal is broad. The zygomatic plate is cut back sharply above, with forward projecting processes on the upper border. The incisive foramina extend to the tooth rows. The palate is broad, reaching behind $M^3/$. The lateral pit is well developed. The pterygoid fossae are unusually deep. There is a prominent coronoid process on the mandible. The upper molars are flat-crowned with long, narrow folds surrounded by a thick layer of enamel.

The skull of *S. hispidus* differs from other species by being relatively long and narrow with a mastoid breadth that is usually less than 46 percent of the basal length. The premaxillae project behind the nasals. The grinding surfaces of the second and third upper and lower molars exhibit an "S" or modified "S" pattern on the enamel folds (not evident on worn teeth).

The skull of the white-throated woodrat, *Neotoma albigula*, is narrow interorbitally with not markedly developed supraorbital ridges and a slender rostrum. The incisive foramina are long, extending level to $M^1/$. The bullae are long and variable. The zygomatic plate is cut back above. The palate ends even with the front of $M^1/$. The braincase is rounded and tapers anteriorly. The coronoid process of the mandible is well developed. The molars are flat and prismatic. $M^1/$ has two outer and two inner folds. $M^2/$ has two outer and one inner fold. $M/_1$ has three inner folds. $M^2/$ has two inner folds. $M/_3$ has one inner fold. $M/_1$ and $M/_2$ have three outer folds. $M/_3$ has one outer and one inner fold.

The skull of *N. albigula* differs from other species in having a relatively broad rostrum. The bullae are of moderate size. The sphenopalatine foramina are larger in this species. There is a depressed region between the orbits as compared to a raised area in the muskrat. The skull is smooth. The dorsal margin of the foramen magnum is deeply concave. The palatine spine is pointed anteriorly. The posterior margin of the palate is concave. The distinctive pattern of the occlusal surface of the grinding molars separates this form from other similar-sized rodents. Being a pack rat the presence of this animal in an archæological site may be a natural intrusive burial of a much later date than the date of the site under investigation.

The skull of the mountain vole, *Microtus montanus*, is angular with a broad braincase and a short rostrum. The supraorbitals usually fuse into a median interorbital crest. The squamosals have more or less well-developed crests. The incisive foramina are constricted posteriorly. The middle upper molar has four closed triangles. A character of the genus is that the lower incisors have roots extending far behind and on the outer side of the molar series. The upper incisors are not grooved. The molars are rootless, with outer and inner re-entrant angles that are approximately equal. The grinding surfaces possess a pattern of sharp-angled enamel folds surrounding dentine. There is considerable geographic variation within this genus.

The skull of the muskrat, *Ondatra zibethicus*, resembles that of *Microtus* but is larger and more massive. It differs from *Neofiber* in the structure of the lower molars. $M/_1$ has six not five triangles between the anterior and posterior loops, and the first one of the six is not closed. The anterior loop is bilobed, with deep re-entrant angles. $M^3/$ has three not two outer salient angles. There are no longitudinal grooves on the anterior surface of the incisor. The infraorbital foramen is small. The skull surface is roughened and the squamosals are enlarged at the expense of the parietals. The posterior border of the palatines terminates in a median spinous process. The roots of the lower incisors are on the outside of the cheek teeth. The interorbital ridge is high and the nasals are broad. The auditory bullae are small. There is a definite constriction between the orbits and the postorbital processes are prominent.

Zapodidae: The zygoma are simple. The bullae are uninflated. The infraorbital foramen is large. The zygomatic plate is broadened and tilted upward. The incisors are sometimes grooved. The cheek teeth are brachydont or hypsodont and are cuspate or laminate. When cuspate the cusps are arranged in three rows; when laminate the folds are pressed closely together.

The skull of the western jumping mouse, *Zapus princeps*, has a moderately swollen frontal region and auditory bullae that are not inflated. The braincase is high and rounded. The anterorbital foramen is large and oval. The zygomata are not widespread but are expanded anteriorly where the malar extends upward to the lachrymal. The zygomatic plate extends nearly horizontal. The upper premolars are small with $M^2/$ smaller than $M^1/$. There are two labial re-entrant folds of equal length on $M^3/$. The anterior cingulum of $M^1/$ is small. The upper premolar is peg-like. The second and third upper molars are nearly alike, the third is reduced and nearly circular. The incisors are deep orange in color and have longitudinal grooves on the outer face of the upper pair.

The species Z. *princeps* differs from others in having a broad, tapered rostrum, and an infraorbital canal less than 1 mm. in diameter (at the bottom of a false infraorbital foramen, five times greater in diameter, which transmits part of the masseter muscle). The upper molars have shallow re-entrant folds. The mandible is weakly developed.

Erethizontidae: Only one species, the porcupine, *Erethizon dorsatum*, is discussed for this family. The skull is compact, broad and heavily constructed. The nasals are short and broad. The frontals are broad with heavy ridges that converge posteriorly forming the sagittal crest. The zygoma is simple and the jugal is deeper anteriorly. The facial region of the skull is short and broad. The bullae are prominent. The upper root of the zygoma is above the anterior base of the zygoma. The heavy skull is somewhat swollen interorbitally. There is no postorbital process. The upper incisors and the anterior portions of the premaxillae project well beyond the tips of the nasals anteriorly. The palate is narrow and short extending posteriorly to the anterior border of the last molar. The jugal element is not supported by the zygomatic process of the maxilla. The infraorbital foramen is extremely large, usually larger than the foramen magnum. The inferior angular process of the mandible is strongly enrolled. The mandible has a low coronoid and angular process. The teeth are rooted and subhypsodont. The upper molars have one persistent internal and one persistent external fold. The incisors are not grooved and are reddish-orange in color on the outer face.

Carnivora

The general carnivore skulls are differentiated by having sharp cusped cheek teeth with carnassials usually present (always the last upper molar and the first lower molar). The braincase is relatively large. The zygomatic arches are strong and wide-spreading. The sagittal and occipital crests are strong. The dental battery consists of incisors for nibbling, canines for tearing, seizing and holding prey, premolars and carnassials for shearing and cutting and molars for crushing and grinding.

Canidae: The canids have small, spatulate, tricuspid incisors. In the dog and wolf the canines are large and strong but more slender and curved in the foxes. All have long and narrow sectoral teeth with sharp cutting edges. The carnassials are exclusively cutting in function. The last molars are much reduced in size.

The problems of determining the differences which separate the dog, *Canis familiaris*, from the wolf, *Canis lupus*, and the coyote, *Canis latrans*, have still not been settled to the satisfaction of those concerned. Many keys have been published describing differences which separate these closely related canids. Some of these can be used to separate the majority of the skulls but there is always the problem of the possible wolf-huskie or dog-coyote hybrid that will fall outside the established "true breed" keys. One such key, long in use and subject to this possibility of error, is that offered by W. E. Howard (1949). This key provides a means of distinguishing between the skulls of the dog and coyote by using the ratio between the palatal width between the upper first premolars and the length of the upper molar tooth row. This is generally a good means of distinguishing between these forms but it is the always present variant that makes it not entirely reliable.

At the present time Miss Barbara Lawrence, of the Museum of Comparative Zoology at Harvard University, is working on an exhaustive statistical study of these forms and when finished it should be a definitive monograph on the subject. After many discussions with Miss Lawrence, over the past several years, I feel that the observations put down here are reliable. In most cases it is a combination of characters rather than single features that will separate the questionable dog-wolf-coyote skulls.

The skulls of the dog, *Canis familiarus*, and the wolf, *canis lupus*, are separated as follows: The large size, big carnassials and canines, well rounded bullae, palate ending anterior to the end of the tooth row, broadly spreading zygomatic arches are all typical of big northern wolves and easily distinguish them from dogs. The main differences occur chiefly in the postpalatal region but to a certain extent these differences also occur in the frontal and zygomatic areas. The lengthening of the skull takes place in this mid-region. As the tooth row does not lengthen correspondingly the palate extends beyond $M^2/$. The tooth row is short compared with the distance from the end of the row to the bulla (the ratio of the length from the posterior margin of the alveolus of $M^2/$ to the depression between the styloid process and the inflation of the bulla to length from C to $M^2/$). In dogs it is 64–80 and in wolves 59–62. The ratios of width across postorbital processes to zygomatic width is greater in dogs. Also, the ratio of the palatal length to zygomatic width is greater in dogs. Wolves usually have a better-developed sagittal crest than that found in dogs. In wolves the sagittal crest is almost horizontal and well drawn out to overhang the occiput, causing the dorsal margin to curve down in this region. Dogs that are large enough to have a strongly projecting sagittal crest have a palate larger than the tooth row as well. Dogs generally have a higher braincase with a noticeable swelling of the frontals or have an elevated forehead. The orbits in dogs tend to be rounder and the bullae are flatter.

The skull of the coyote, *Canis latrans*, has no pit at the base of the post-orbital as in the fox. The skull exceeds 150 mm. in length (that of the wolf exceeds 225 mm.). There is a prominent sagittal crest (this is also at times present in the dog). The tips of the upper canine teeth usually fall below a line drawn through the anterior mental foramina in the lower jaw when articulated. In the dog and red wolf the tips of the upper canines usually fall well above this line. There is a more noticeable swelling in the frontal area in the dog. The upper incisors are prominently lobed in the genus *Canis*. The upper canines in the coyote measure less than 11 mm. at the base as compared with a measurement of 12 mm. or more in the wolf. The premolars are relatively narrow. The zygomatic breadth is usually less than 105 mm. The width of the rostrum measured across the bases of the upper canines is less than 1 ¾ the anteroposterior extent of the bulla, whereas in the wolf the width of the rostrum is more than 1 ¾ the anteroposterior extent of the bulla.

The skull of the red fox, *Vulpes fulva*, has temporal crests that are not prominent. When present they are separated by a space less than 10 mm. (more than 10 mm. in the gray fox). The postorbital process is thin, with a shallower pit than that found in the gray fox (no pit in *Canis*). There is no step present below the angle of the lower jaw. There are slight lobes on the upper incisors. The width $M^1/$ is noticeably greater than the length, resulting in a narrower tooth. The length and width are more nearly equal in the gray fox resulting in a more square tooth. $M^2/$ is noticeably smaller than $M^1/$. $M^1/$ and $M^2/$ are nearly the same size in the gray fox.

The skull of the kit fox, *Vulpes macrotis*, is about one-quarter less in all measurements than in the red fox. The length of the auditory bullae (measured from ventral-most union with the paroccipital process to carotid foramen) is more, rather than less, the greatest breadth of the rostrum over the canines; and more, rather than less, the distance between the posterior border of $P^4/$ and the anterior border of $P^3/$. The distance between the orbit and the anterior opening of the infraorbital canal is less, rather than more, the height of the foramen magnum. The anterior palatine foramina extend posteriorly past rather than to the middle of the upper canines. The premolars are widely spaced. The auditory bullae are close together. There are no lobes on the upper incisors.

The skull of the gray fox, *Urocyon cinereoargenteus*, has a relatively shorter rostrum than that found in the red fox. It has a distinctive lyrate pattern of the temporal ridges on top of the skull and a longitudinal pit at the base of the postorbital process (shallower in the red fox, absent in *Canis*). The temporal ridges are separated by a space of more than 10 mm. A noticeable step is present in the margin of the lower jaw below the angle. There is a supplementary tubercle on $M^1/$. The length and width of $M^1/$ are more nearly equal. The width of $M^1/$ in the red fox is noticeably greater, resulting in a narrower, less square tooth than that found in the gray fox. $M^1/$ and $M^2/$ are nearly the same size. $M^2/$ is noticeably smaller in the red fox.

Ursidae: In the ursid skull there is an alisphenoid canal present. The auditory bullae are depressed, and consist of a single bone (tympanic) which is readily detached from the cranium in young animals, and is more or less triangular in shape. The molars have an elongated and crenulated grinding surface. The carnassial has lost much of its shearing function but no hypocone has developed. The protocone has migrated backward. The incisors are tricuspid and single-rooted. The canines are strong, with long pointed roots and sharply pointed crowns. The premolars are small, pointed teeth. The molars are multicuspid and are low-crowned grinders. The fourth upper premolar (carnassial) is without a third or inner root.

The skull of the black bear, *Euarctos americanus*, has uninflated tympanic bullae. The alisphenoid canal is present. The molars have broad, flat tubercular crowns. The inferior lip of the long auditory meatus is prolonged. There is a large paroccipital process of the exoccipital bone, standing free and not applied to the bullae. The condylid and glenoid foramina are distinct. The dorsal profile of the skull is flat or gently rounded, not sharply depressed in the frontal region. There are large nares and exposed turbinal bones. The postorbital processes are well developed. The carnassials are not developed.

The skull of *E. americanus* has the specific characters of the combined length of $M^1/$ and $M^2/$ never less than the palatal width. $P^4/$ is without a medial accessory cusp. The last molar is larger than the one in front of it. The bony palate extends posterior to the last upper molar.

The skull of the grizzly bear, *Ursus horribilis*, is long and narrow with a long rostrum with a concave facial contour. The skull is generally large and massive with the temporal ridges or depressions meeting anterior to the posterior ends of the frontals and varying somewhat, being sometimes very well marked ridges and at times being merely indicated. The teeth are massive with $M^1/$ having one or more cusplets set medially in the valley between the metacone and the protocone. $P^4/$ has accessory cusps. $M^2/$ is broadest at the anterior end.

Procyonidae: The alisphenoid canal is present in the skull of the procyonids. There are two well-developed molars with adequate grinding surfaces. The carnassials have lost their shearing function, and the upper ones have developed a hypocone. There are three cusps along the outer margin of the upper carnassial, which has a broad, bicuspid inner tubercle, forming a nearly quadrate crown.

The skull of the cacomistle, *Bassariscus astutus*, is elongate with the postorbital processes well developed from the frontals and moderately developed from the zygomata. The temporal crests unite, if at all, only posteriorly to form the sagittal crest. The palate ends about at the posterior plane of $M^2/$. The auditory bullae are moderately inflated. The braincase is flattened but expanded laterally. The postorbital processes are prominent. The zygomata are slender. The molars and premolars have pointed cusps. The incisors have secondary lobes. The canines are rounded. There are four upper and four lower premolars. $P^1/$ and $P^3/$ are unicuspid. $P^4/$ is sectorial with a well-developed blade. $M^1/$ is variable, subquadrate to triangular, and broader than long. The mandible is slender.

The skull of *B. astutus* differs from other procyonids in having high, sharp ridges connecting the cusps of the molariform teeth and in having the upper carnassial irregular in outline.

The skull of the raccoon, *Procyon lotor*, has a rounded dorsal profile, a broad braincase and a broad rostrum. The bullae are inflated on the inner side. The molar ends at $M^1/$ and the palate extends behind the posterior for a distance of more than one-fourth the total length of the palate. The internal nares are not divided by a septum. The molars are flat-crowned for crushing. The upper molariform teeth have sharp coniform cusps. There are five principal cusps on the last upper molar. This tooth is not a carnassial. $P^4/$ has five tubercles. $I/_3$ usually has an independent cusp. The large grooved canines have an oval cross section.

The skull of the coati, *Nasua narica*, is long and narrow having a long rostrum which is compressed laterally. The interorbital constriction is slight. The sagittal crest is well developed. The palate is flat or hollowed posteriorly with distinct lateral grooves that extend far beyond the plane of the last molars. The palatine bones are deeply notched on each side and have long processes which extend laterally to the maxillary tuberosities. The vomer is usually attached to the palatal bones. The foramen ovale is small. The bullae are small and smoothly rounded, rising at a sharp angle from the tube of the external auditory meatus. The skull rises above the rostrum. The internal nares are divided by a septum. The ascending ramus of the mandible is relatively small and low. $I^3/$ is separated from $I^2/$ by a diastema. The lower incisors protrude forward. The canines are compressed laterally and are grooved on the inner surface.

Mustelidae: The mustelids have a low, elongated skull with orbits far forward and with a comparatively short muzzle. The facial region is shortened. The tympanic bullae are generally formed by the tympanic bone alone. There is a long canal running fore and aft beneath the bullae for the carotid artery. In most mustelids the upper molar has a characteristic expansion of the inner portion with a "waist" between the two parts of the tooth. The incisors are small. The canines are slender and sharp. $P/_1$, if present, is small. The premolars have one main cusp. $P^4/$, the carnassial, may have a lingual subsidiary cusp. $M^2/$ is a grinding tooth and $M/_2$ is vestigial. There is no alisphenoid canal present.

The skull of the marten, *Martes americana*, is smooth and dorsally rounded, not angular, and the frontal region is slightly depressed with a slight facial angle. The palate ends behind the last upper molars. The infraorbital foramen is less than 6 mm. The tympanic bullae are moderately inflated and not in close contact with the paroccipital process. In $M/_1$ the trigonid is longer than the talonid.

The skull of the long-tailed weasel, *Mustela frenata*, has the following mustelid characters: The facial angle is slight, and the tympanic bullae are greatly inflated, elongated and cancellous in structure. The paroccipital processes are closely oppressed to the bullae. The palate ends behind the upper molars. The rostrum is short. There is a dumbbell-shaped upper molar.

M. frenata has a skull with the following specific characters: The rostrum is quite long for a mustelid, the tympanic bullae are well inflated and project below the squamosal bone anterior to the bullae. The skull of the adult has a well-developed sagittal crest. The postorbital process is well developed. There is a pronounced postorbital constriction. The braincase is triangular-shaped. The lambdoidal crests are well developed. The zygomatic arches are widespread and highly arched. The bullae are large, rounded and are about twice as long as wide. The postglenoid length of the skull is less than 47 percent of the condylobasalar length. The length of the two lower mandibles in normal position is more than the postglenoid length (glenoid fossa to exoccipital condyle).

The skull of the mink, *Mustela vison*, has the following characters which separate it from the mustelids: It differs from the long-tailed weasel in its larger size and more nearly flat bullae. The skull is moderately strong and somewhat flattened with a short broad rostrum and evenly spreading zygomatic arches. The lambdoidal crests are well developed in the adult, and usually extend posteriorly almost as far as the posterior border of the condyle thus almost obscuring the foramen magnum in superior view. The bullae are longer than wide. The bony palate extends posteriorly beyond the back molars, the extension being nearly as great as the depth of the palatal notch. There are four molariform teeth on each side. The last upper molar is dumbbell-shaped and is smaller than the preceding tooth.

The skull of the black-footed ferret, *Mustela nigripes*, has the following specific characters: The skull is large and massive, very broad behind the orbits, and deeply constricted behind the postorbital processes. The postorbital processes are well developed. The bullae are flattened obliquely on the outer side. The distance between the upper canines is more than the width of the basioccipital as measured between the foramina situated along medial sides of the tympanic bullae.

The skull of the badger, *Taxidea taxus*, is large and rugose with a depressed occiput and a steep facial angle. The skull is wedge-shaped, broad posteriorly and narrow anteriorly with an angular braincase. The bullae are highly inflated and not in contact with the paroccipital process. The palate extends posteriorly beyond the plane of the upper molars. The occipital region is wide and truncate. The posterior part of the skull is almost as broad as the zygomatic width. There is a large infraorbital foramen, triangular in anterior view. The lambdoidal crest is greatly developed but the sagittal crest is little developed. $M^1/$ is triangular, often with cusps arranged in rows that are transverse to the long axis of the skull. The dental formula separates this animal from other similar-sized forms. The jaws are so strongly articulated that they commonly remain in position even after the soft connecting tissue has been removed.

The skull of the striped skunk, *Mephitis mephitis*, has these generic characters: The skull is arched and highest in the frontal region with the rostrum below the plane of the frontals. The infraorbital canal opens above the posterior half of the fourth upper premolar. The mastoid bullae are relatively uninflated. The posterior margin of the palate is nearly on a line with the posterior border of the upper molars. The inferior margin of the mandible is curved. The angle of the mandible is developed as a flattened face in a vertical plane, producing a "step" or concavity in the inferior margin. The coronoid process is high and vertically inclined. The anterior and posterior diameters of $M^1/$ are equal to (or more than) the outside length of $P^4/$.

The skull of the striped skunk differs from other mustelids in having small tympanic bullae. The anterior palatine foramen are usually small and narrow and there is a broad interpterygoid fossa. The palate ends squarely without a median notch or spine. The last upper molar is squarish in outline.

Mephitis is separated from the hog-nosed skunk *Conepatus* by having the width of $M/_1$ slightly less than one-half the length while in *Conepatus* the width is slightly greater than one-half the length. The $P^4/$ in the striped skunk is large, conical and separated from the shearing blade only by a narrow crease, never developing a subsidiary cuspule. In the hog-nosed skunk the protocone of $P^4/$ is small, laterally compressed, well separated from the shearing blade by a broad valley and tending to develop a subsidiary cuspule. In the striped skunk $M^1/$ is wider than long, the lingual half of the crown is not displaced posterad. The metacone is the most posterad portion of the tooth. In the hog-nosed skunk $M^1/$ is longer than wide or occasionally subequidimensional, the lingual half of the crown is displaced posterad so that the hypocone is the most posterad portion of the tooth.

The skull of the hog-nosed skunk *Conepatus leuconotus*, is deepest in the temporal region. There is a marked interorbital constriction and relatively slender but widespread zygoma. The interpterygoid space is relatively narrow. The mastoid bullae are hardly inflated. The posterior margin of the palate ends behind the molars. The inferior margin of the mandible is longitudinally convex. The anteroposterior and transverse diameters of $M^1/$ are each more than the outside length of $P^4/$.

The skull of the river otter, *Lutra canadensis*, is strongly flattened with a broad braincase which is long and arched. The muzzle is short and the interorbital breadth is less than the width of the muzzle. The infraorbital foramen is large as, or larger than, the alveolus of the canine. The orbit is open and the postorbital process varies from large to rudimentary. The alisphenoid canal is absent. The opening of the external auditory meatus is large and the paroccipital process is not in contact with the bulla. The rostrum is broader than it is long. The tympanic bullae are flattened. The palate extends beyond the last molar. The infraorbital foramen is oval in anterior view and not triangular as in the badger. The second lower incisor is large and is crowded behind the others. The upper molar is squarish and is broader than long.

Felidae: The felid skull is short and rounded with wide, strong zygomatic arches. The facial portion is especially short and broad. The cranium is rounded and the auditory bullae are large, rounded and smooth. There is no alisphenoid canal. The canines are powerful. The carnassials are well-developed shearing blades. The cats are almost entirely carniverous and there is no grinding surface left in the dental battery. Only a vestigial first upper molar remains behind the carnassial. The incisors are small and close set. $P^1/$ is missing. $P^2/$ and $P^3/$ are small. $P^4/$ is enormous (carnassial). $M^1/$ is a vestige. $P/_1$ and $P/_2$ are absent. $P/_3$ and $P/_4$ are cutting teeth. $M/_1$ is a large carnassial. The canines are recurved with trenchant edges and sharp points. The carnassials are three-lobed and the lower are bilobed.

The skull of the jaguar, *Felis onca*, is highly arched in the frontal region with a short rostrum. The face is flattened. The carotid canal is short or absent. The alisphenoid canal is absent and the tympanic bullae are large. The palate ends posterior to the molars. Besides the above generic characters the jaguar has these specific diagnostic characters: The bregmatic processes of the parietals do not extend anteromedially over the frontals, approaching or reaching the temporal crest and fusing in the old animals as does the puma. The dorsal profile of the skull has a sagittal concavity, and the skull in general is relatively elongate.

The skull of the ocelot, *Felis pardalis*, is highly arched and the nasals do not extend anteriorally to the plane of the anterior edge of the palatine foramina. The palatine foramina are visible in dorsal aspect. The rostrum is abruptly truncated.

The mandible of the ocelot can be distinguished from that of the bobcat by the following: In the ocelot the anterior mental foramen lies below the posterior half of the diastema or below the anterior root of $P/_3$. The posterior foramen lies below the posterior half of $P/_3$ or still farther posterad. In most specimens of bobcat the anterior foramen lies below the midpoint of the diastema, and the posterior foramen lies below the midpoint of the anterior half of $P/_3$. In the ocelot the diastema between the lower canine and $P/_3$ is shorter than the anteroposterior diameter of $P/_3$, or equal to it in a few cases. In the bobcat the diastema is almost invariably longer than $P/_3$ or occasionally equal to it.

The skull of the puma, *Felis concolor*, has the bregmatic processes of the parietals extending anteromedially over the frontals, approaching or reaching the temporal crest and fusing with the frontals in old individuals (not so in the jaguar). The dorsal profile of the skull is convex, without a sagittal concavity (opposite in the jaguar). The nasal bones are particularly wide anteriorly. The mandible is deep and short. There are three instead of two premolars in the upper jaw (there are two in the bobcat). There are four cheek teeth above, the first and last are small and single-rooted.

The skull of the yagouaroundi, *Felis yagouaroundi*, is only slightly arched, with the nasals extending anteriorly beyond the plane of the anterior edge of the palatine foramina. The palatine foramina are not visible in the dorsal aspect. The skull is elongate and the cranium is compressed laterally. The rostrum is sharply elevated and the bullae are large and constricted laterally.

The skull of the bobcat, *Lynx rufus*, has the generic characters as follows: The skull is as in *Felis* but with the nasal branch of the premaxillae slender and gradually attenuated. The postorbital processes are thinner and less depressed and sharper. The notching of the suborbital edge of the palate is shallower. Specific characters are as follows: The face is short and the skull is rounded. The skull is robust and the sagittal crests are lyrate. The lambdoidal crests and inion are well developed. The tympanic bullae are large. The postorbital processes are prominent. The anterior condyloid foramen is confluent with the foramen lacerium posterius. The palatal exposure of the presphenoid is strap-shaped or slightly triangular but in the canada lynx the presphenoid is broadly flask-shaped. There are two premolars in the upper jaw as compared with three in the puma. Compare with the description of the skull of the ocelot, *Felis pardalis*.

Perissodactyla

Equidae: Only one species is considered in this study, the horse, *Equus caballus*. The facial part of the skull is long with the nasals extending freely over the external nares. In the burro the orbits extend laterally to more of a degree than in the horse. The occiput overhangs the basicranium, making more of an acute angle in the burro than that which is present in the horse. The teeth are truly diagnostic for this animal, being noticeably different from all other similar-sized mammals. The molars of *Equus* tend to be square, with high-crowned grinding surfaces capable of taking considerable wear before being worn to the root area. The grinding surface usually consists of a series of curved ridges of varying pattern. There is usually very little difference between the premolars and the molars, the entire series generally being designated as cheek teeth. There is a large diastema between the first premolar and the incisors and canines (when present). The incisors are chisel-like cropping organs. The canines are generally attributed only to the stallions but are occasionally found in the mares (Pope, 1934). The cheek teeth protrude from the alveoli to keep pace with the wear of the grinding surfaces. The teeth are adapted exclusively for grazing where much abrasive material is taken into the mouth along with the food. The crimped, file-like enamel pattern is well designed to crush and grind up a fibrous diet.

Much has been published concerning the different species of *Equus* that were present in sub-Recent and Pleistocene times. Occasionally these teeth turn up in archæological digs, perhaps as a part of a medicine bag or having been picked up as a curiosity by a pre-Columbian Indian much as modern man does to this day. The descriptions of some species of Pleistocene horses were based on isolated teeth or at best on incomplete dentitions. When, for purposes of publication, these teeth were being compared with the teeth of other reported species of the genus *Equus*, they were noted as differing by being more curved in outline, or by being more robust, or by other vague morphological characters. In reality all that this means is that the specimens being described differed in this manner from the few specimens with which they were compared, in many cases only the types of other species. Actually a larger series, where available, will show considerable variation, due to age or to the individual, and of the same degree as that stated as being a valid character for a new species of *Equus*. Careful study, aided by the accumulated knowledge of the past few decades, has synonomized other similar groups of Pleistocene mammals whose early generic categories contained many species described from fragmentary material. This is particularly true of the tapirs and peccaries of the Pleistocene or sub-Recent deposits.

Several experts on the evolution of the horse have commented on this Recent-Pleistocene horse problem and a few of their comments are worthy of note. Professor F. B. Loomis (1926) stated, "The number of species found in this early phase is bewildering, over twenty being already described. Several are based on teeth alone. . . . The English have been very conservative in reporting the Pleistocene horses from Great Britain, calling them all *Equus caballus*." Dr. J. E. Ewart (1904) reports, "The great French naturalist Cuvier believed not only that all living horses belonged to one species (the *Equus caballus* of Linnaeus), but also that there was no specific difference between living breeds and fossil horses of the Pleistocene period. . . . Except in size I have been unable to discover any difference between the skeleton and teeth of the Celtic pony and those of the small horse of the Brighton Pleistocene. . . . it is extremely probable that some of the prehistoric (pre-Glacial) varieties have persisted almost unaltered to the present day." Professor R. S. Lull (1931) states, "*Equus*, the modern horse, first appears in the Early Pleistocene beds of Eurasia and North America. In point of tooth structure, one of these (of 27 recorded forms), *Equus fraternus*, resembles closely the modern horse, *E. caballus*." Dr. C. E. Ray (1957) based his age for the pre-Columbian *Equus* from Yucatan on the depth of burial and the degree of mineralization.

In view of my own comparisons of horses and what has been said by specialists in early horses, one must be extremely careful in assigning an age to *Equus* remains when they are isolated and consist of fragmentary teeth or scraps. It is entirely possible to have Pleistocene horse teeth present in a site and have them attributed to a post-Columbian form.

ARTIODACTYLA

There is considerable variation among the skulls of the families of artiodactyls and these will be treated in detail under the family and generic headings.

Tayassuidae: The skull of these forms has an elevated and backward sloping occipital crest. The palate is long, narrow and extends behind the last molar. The paroccipital processes are long. The muzzle is long and narrow with a small nasal opening at its extremity.

The small postorbital processes do not meet the zygomata, so the orbits are open posteriorly. The lachrymal extends beyond the orbit for a considerable distance on the cheek. The domestic pig has canines that curve outward and upward. The wild peccaries have canines that remain in a "normal" canine position. The upper incisors are short and stout with curved roots. The lower incisors are long and slender with straight roots. The lower canine's distal surface rubs with every movement of the jaws against the mesial surface of the upper canines, forming an occlusal facet, having a sharp point and cutting edge. The premolars and molars have a multiplicity of tubercles (not unlike a human molar), the whole combination producing an efficient grinding surface of enamel folds and dentine.

The skull of the domestic pig, *Sus scrofa*, has a very high occipital crest and long and narrow nasals. The palate is long and narrow and extends back beyond the plane of the last molars. The upper canines are directed outward and upward.

The skull of the wild peccary, *Pecari tajacu*, has a rounded profile when viewed laterally. The maxilla are not laterally expanded above the tooth row. A distinct ridge is present on the palate from C to $P^2/$. The rostrum is narrow with no flattening of the sides and is divided by a zygomatic ridge. The infraorbital canals are rounded. The upper canines are directed downward (not outward and upward as in the domestic pig). The upper canines have sharp cutting edges on the posterior face. The molars have well-developed cingula. The cusps are not closely connected by intermediate cusplets. There is a rounded notch on the ventral margin of the mandible, anterior to the angle.

Cervidae: The skull of the cervids have bones that are noticeably thin and brittle. The parietals form the greater part of the cranial vault, but the frontals contribute to the narrow anterior portion. Large squamosals are present. The orbits are large and completely surrounded by bone. There is often an unossified space on the cheek, between the frontal, nasal, lachrymal and maxillary bones and anterior to the orbit. This is the lachrymal fossa (but has nothing to do with the tears, housing instead a scent gland). Antlers are present, generally in the males. They are true bony growths that arise from short bony processes of the frontal bones and are shed and developed anew each year. In most deer the orifice of the lachrymal canal is double, and situated on the margin of the orbit, whereas in most hollow-horned ruminants it is single and situated well within the margin. The nose is directed forward in relation to the basicranial axis in the deer (directed downward in the sheep and goats). The upper canines and incisors are missing. There is a long diastema anterior to the cheek teeth. The cheek teeth consist of molariform molars and premolars. These are selenodont, with low crowns. The premolars have a single pair of cresentric cusps. $M/_3$ has in addition a third heel-like cusp. The lower cheek teeth are somewhat narrower than the upper. The lower incisors in life are opposed by a hard fleshy pad borne on the premaxilla, against which the food is held by the lower incisors and pulled apart. The cheek teeth have distinct roots (are not pillared).

Because the skull of the elk, *Cervus canadensis*, is long and narrow, size alone will narrow down the possibilities of identification. The large branched antlers when present are also diagnostic. The posterior narial cavity is not completely divided by the vomer. The frontal ridges are not prominent. The lachrymal fossa is moderately large. The cheek teeth are moderate and short-crowned. Wide maxillary canines are present in both sexes. The lower incisors are distinctly, but not excessively, differentiated in size and form (fig. 2).

The skull of the mule deer, *Odocoileus hemionus*, has small lachrymal pits and the lachrymal fossae are shallow with the duct on the rim of the orbit. The lachrymal bone is not connected with the nasal bone. The nasal cavity is not separated by the median vomer. The premaxillae do not reach the nasals. Canines are usually absent. Antlers are present in the male only. The antlers have equal dichotomous branches. There are no upper incisors.

The skull of the white-tailed deer, *Odocoileus virginianus*, has a space of five-eighths of an inch between the orbit and duct. The width of the auditory bulla is equal to the length of the bony tube which leads to the meatus. There are no upper incisors. The antlers are present in the male only and consist of a main beam curving sharply out and forward and all subsidiary points except the basal emerge from the dorsal surface of the main beam.

The skull of the moose, *Alces americana*, has distinctive palmate antlers, present only in the male. The length of the nasals is less than one-half of the distance from their anterior tips to the anterior tips of the premaxillaries. The premaxillary region is greatly lengthened. The nasals are short and the nasal aperture is large. The vomer is low, not dividing the posterior nares into chambers. There is a wide lachrymal vacuity and a well-developed lachrymal pit. The maxillary canines are usually absent. There are no upper incisors and the lower ones are little differentiated. As with the elk, size alone will differentiate the moose skull from most of the cervids.

Antilocapridae: Only one species is considered.

The skull of the prong-horned antelope, *Antilocapra americana*, has frontal sinuses that open to the outside through two large longitudinal fossae located in the dorsal surface of the frontal bones. The orbits are large, and situated ventral to the horn cores. The lachrymal bone does not articulate with the nasal bone. The profile of the rostrum is convex (compare with deer). There are horns in both sexes. These are deciduous with unbranched bony horn cores, flattened laterally and sharp anteriorly. There are pillared cheek teeth (as compared to deer) and no upper incisors.

Bovidae: The bovid skull has the lachrymal bone usually in contact with the nasal. The lachrymal canal usually has one opening inside the orbit. Horns when present are hollow, unbranched, nondeciduous, and the epidermal covering is renewed by growth from the base. The cheek teeth are selenodont but may be either brachydont or hypsodont. In typical sheep, the basioccipital of the skull is wider in front than it is behind, with the anterior pair of tubercles widely separated and much larger than the posterior pair. The sheep and goats have the nose directed downward in relation to the axis of the basicranium instead of outward as in the deer.

The skull of the domestic goat, *Capra hirca*, has the lachrymal bone only slightly concave, not forming a pit directly in front of the orbit. Horns when present are nearly parallel and curve backward. There are grooves on top of the skull from the inner side of the base of each horn or knob, forming a "V" or "U" on the forehead. There is no antorbital pit. The premaxillaries are deeply wedged between the nasals and the maxillaries. The cheek teeth are pillared, and there are no upper incisors.

The skull of the mountain goat, *Oreamnos americanus*, has the lachrymal depression absent, and there are horns in both sexes, unbranched. The vertical ridges on the labial face of the premolars are prominent. The top of the cranium is scarcely elevated above the top of the orbit. The braincase is arched behind the horns and prolonged anteriorly, not descending steeply to the lambdoidal ridge. The skull is elongate and narrows abruptly from orbits to top of rostrum. The premaxillaries are expanded at their tips, and their nasal processes are widely separated from the nasals. There are no upper incisors.

The skull of the domestic sheep, *Ovis aries*, has a distinctly concave lachrymal bone, forming a deep pit in front of the orbit. Horns, when present, are curved decidedly outward and downward. There is no groove on top of the skull forming a "V" or "U" on the forehead. The premaxillaries are touching or nearly touching the nasals. Cheek teeth are pillared, and no upper incisors are present.

The skull of the mountain or big-horned sheep, *Ovis canadensis*, has the suborbital gland and lachrymal fossa usually present, but they are generally small. The basioccipital of the skull is wider anteriorly than posteriorly, with the anterior pair of tubercles widely separated and much larger than posterior. The premaxillaries touch or nearly touch the nasals. The horns of the male are either spiraled with tips directed outward or bent in an arc with the tips pointing either forward or toward each other behind the head. The coronal suture projects forward in an angle. The infraorbital foramen is small with a well-defined rim about equal to the length of the $P^4/$. The upper ends of the premaxillae are not usually wedged between the nasals and the maxillae. The lambdoidal suture forms more or less of a straight line. The labial sides of the premolars have prominent pillared enamel ridges. No upper incisors are present.

Figures 4 to 44

Skull drawings of the cited forms. Differentiating characters are indicated by notes and arrows or by heavy dashed lines. Elements of the skull, processes and articulations common to all forms, are labeled with straight lines without arrows.

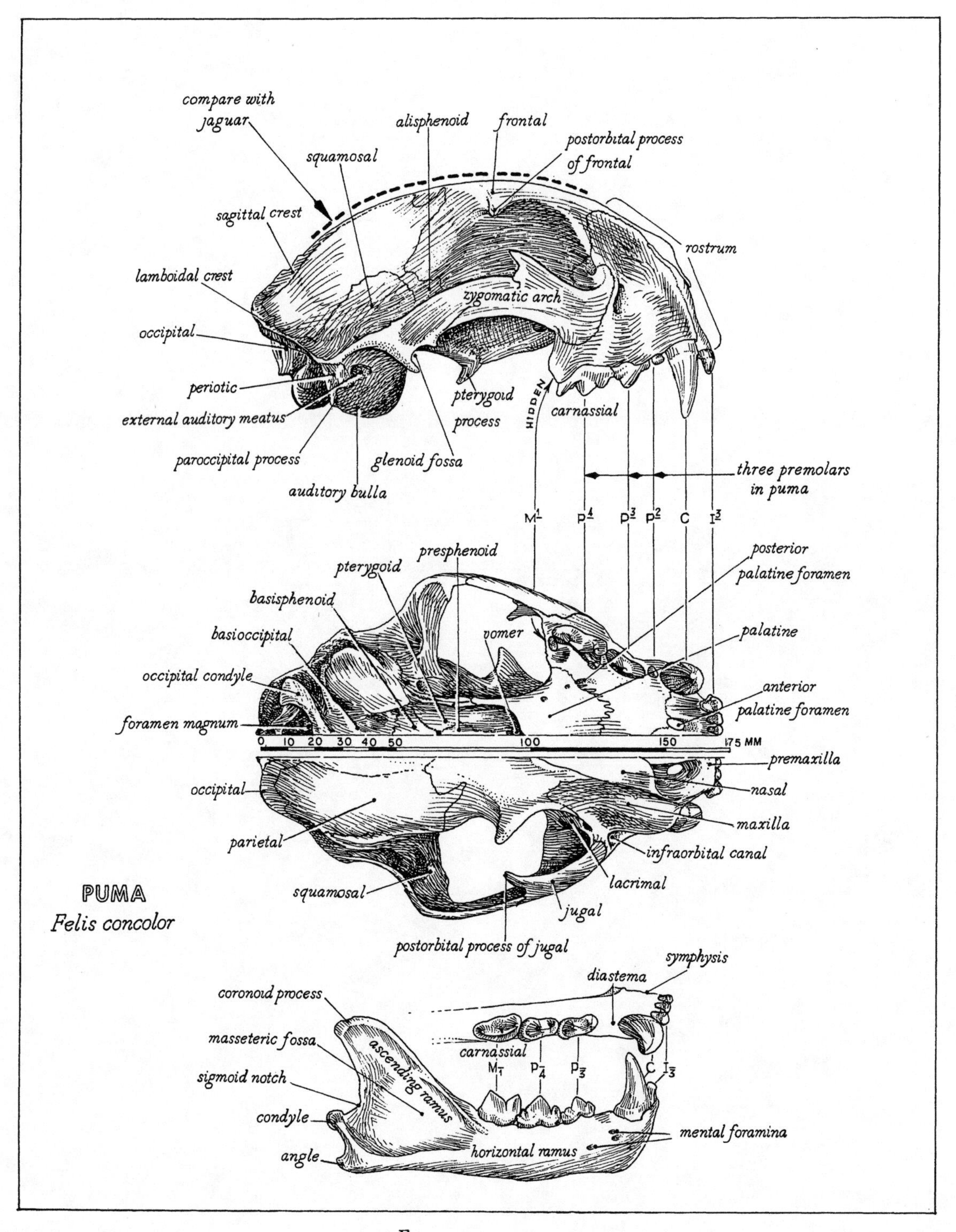

compare with jaguar
alisphenoid
frontal
postorbital process of frontal
squamosal
sagittal crest
rostrum
lamboidal crest
zygomatic arch
occipital
periotic
pterygoid process
HIDDEN
carnassial
external auditory meatus
paroccipital process
glenoid fossa
auditory bulla
three premolars in puma
M1 P4 P3 P2 C I3
presphenoid
pterygoid
posterior palatine foramen
basisphenoid
basioccipital
vomer
palatine
occipital condyle
anterior palatine foramen
foramen magnum
0 10 20 30 40 50 100 150 175 MM
premaxilla
occipital
nasal
maxilla
parietal
infraorbital canal
PUMA
Felis concolor
squamosal
lacrimal
jugal
postorbital process of jugal
symphysis
diastema
coronoid process
masseteric fossa
ascending ramus
carnassial
M1 P4 P3 C I3
sigmoid notch
condyle
mental foramina
angle
horizontal ramus

FIGURE 4

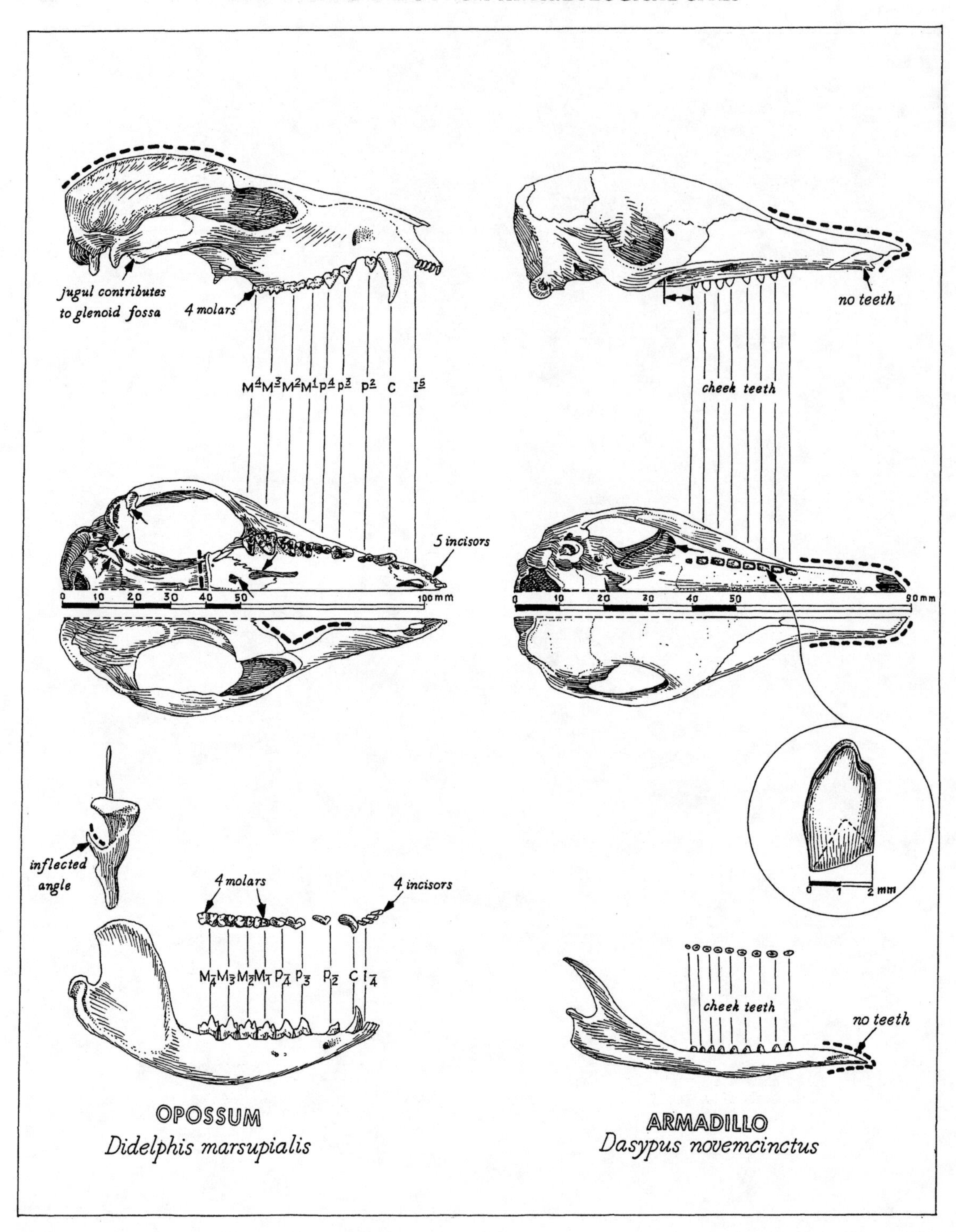
jugul contributes to glenoid fossa
4 molars
cheek teeth
no teeth
5 incisors
100 mm
90 mm
inflected angle
4 molars
4 incisors
cheek teeth
no teeth
2 mm
OPOSSUM
Didelphis marsupialis
ARMADILLO
Dasypus novemcinctus

FIGURE 5

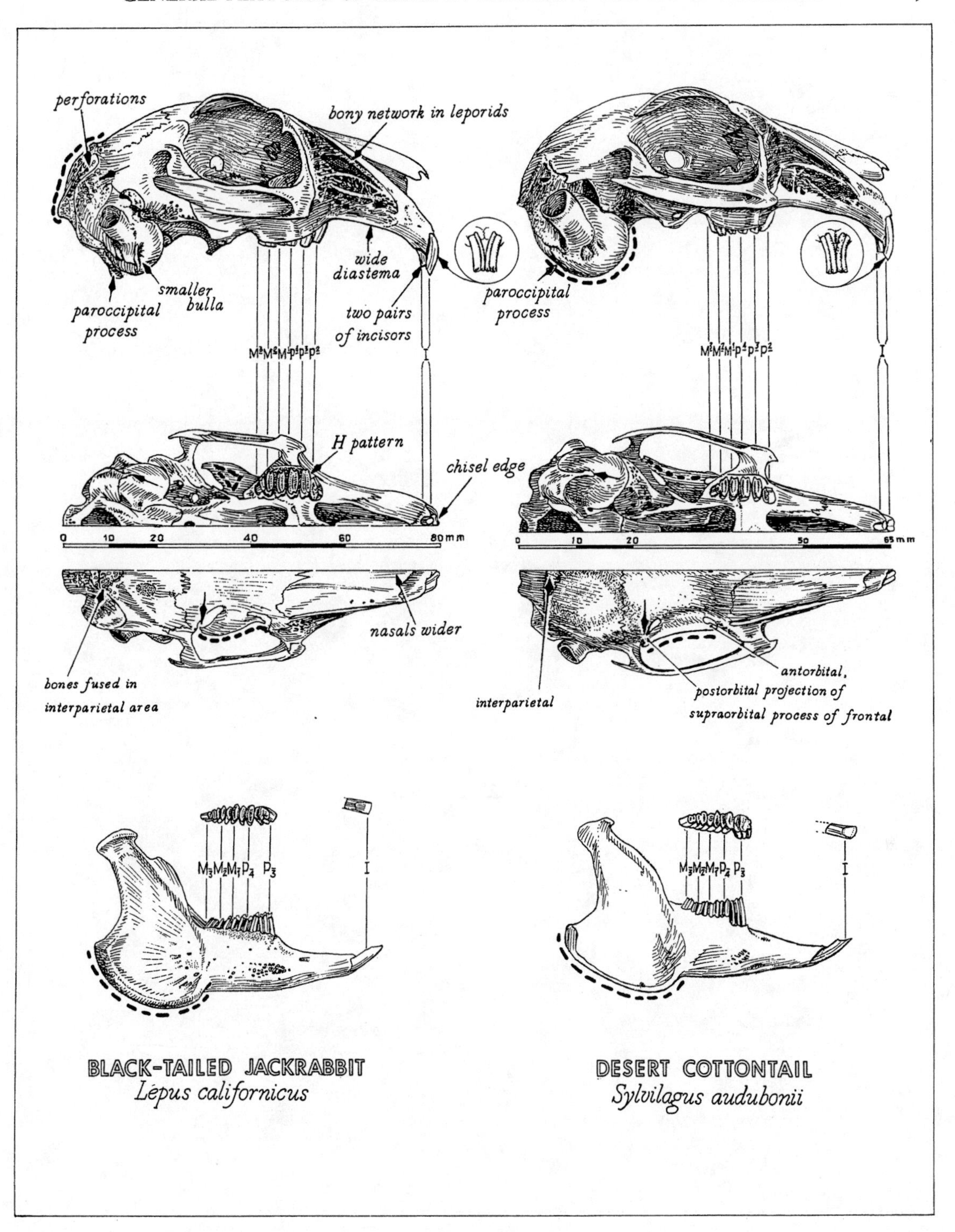
perforations
bony network in leporids
wide
diastema
smaller
bulla
paroccipital
process
two pairs
of incisors
paroccipital
process
H pattern
chisel edge
0
10
20
40
60
80 mm
0
10
20
50
65 mm
nasals wider
bones fused in
interparietal area
interparietal
antorbital,
postorbital projection of
supraorbital process of frontal
I
I
BLACK-TAILED JACKRABBIT
Lepus californicus
DESERT COTTONTAIL
Sylvilagus audubonii

FIGURE 6

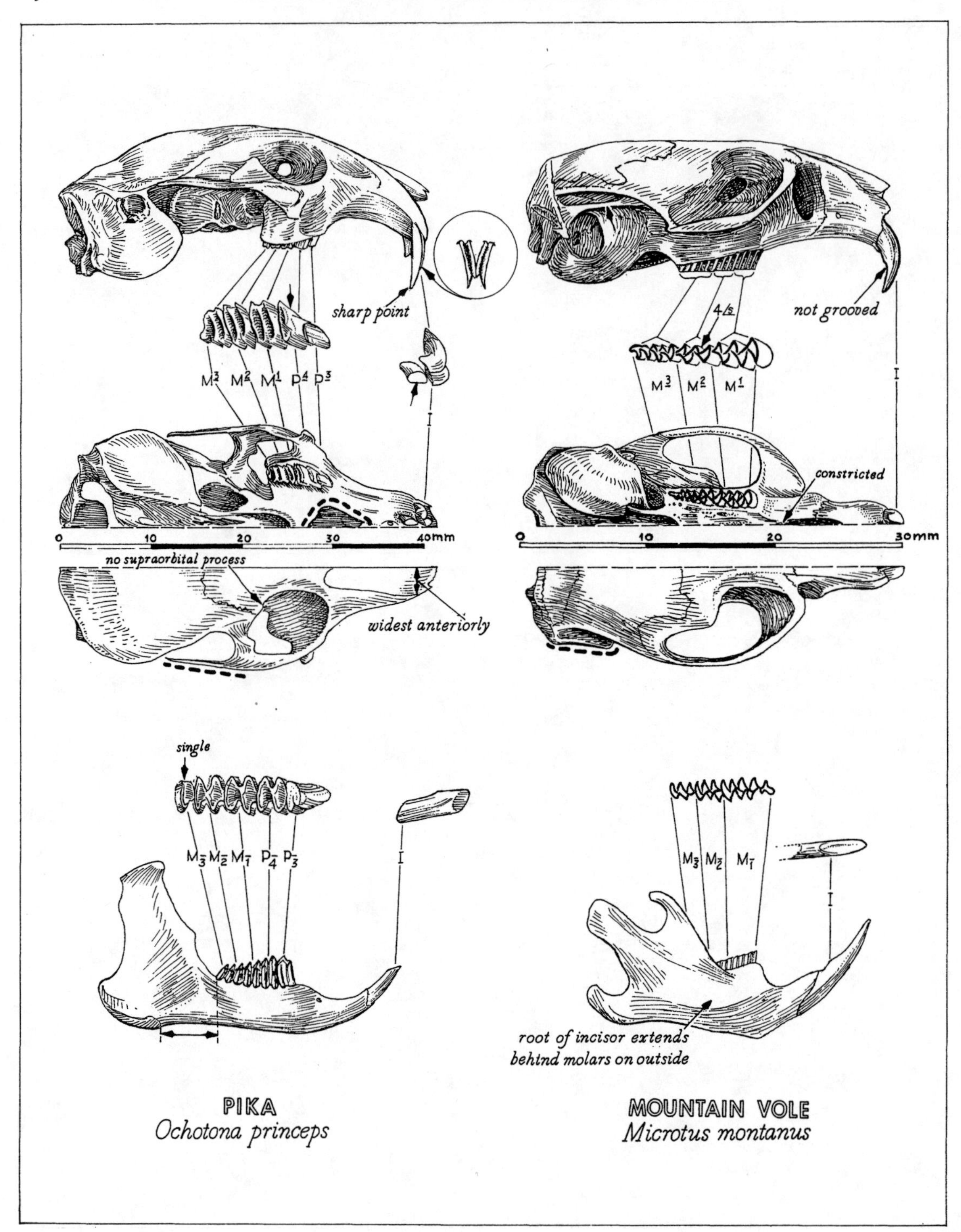
sharp point
M3 M2 M1 P4 P3
I
no supraorbital process
widest anteriorly
0 10 20 30 40mm
not grooved
4/5
M3 M2 M1
I
constricted
0 10 20 30mm
single
M3 M2 M1 P4 P3
I
M3 M2 M1
I
root of incisor extends
behind molars on outside
PIKA
Ochotona princeps
MOUNTAIN VOLE
Microtus montanus

FIGURE 7

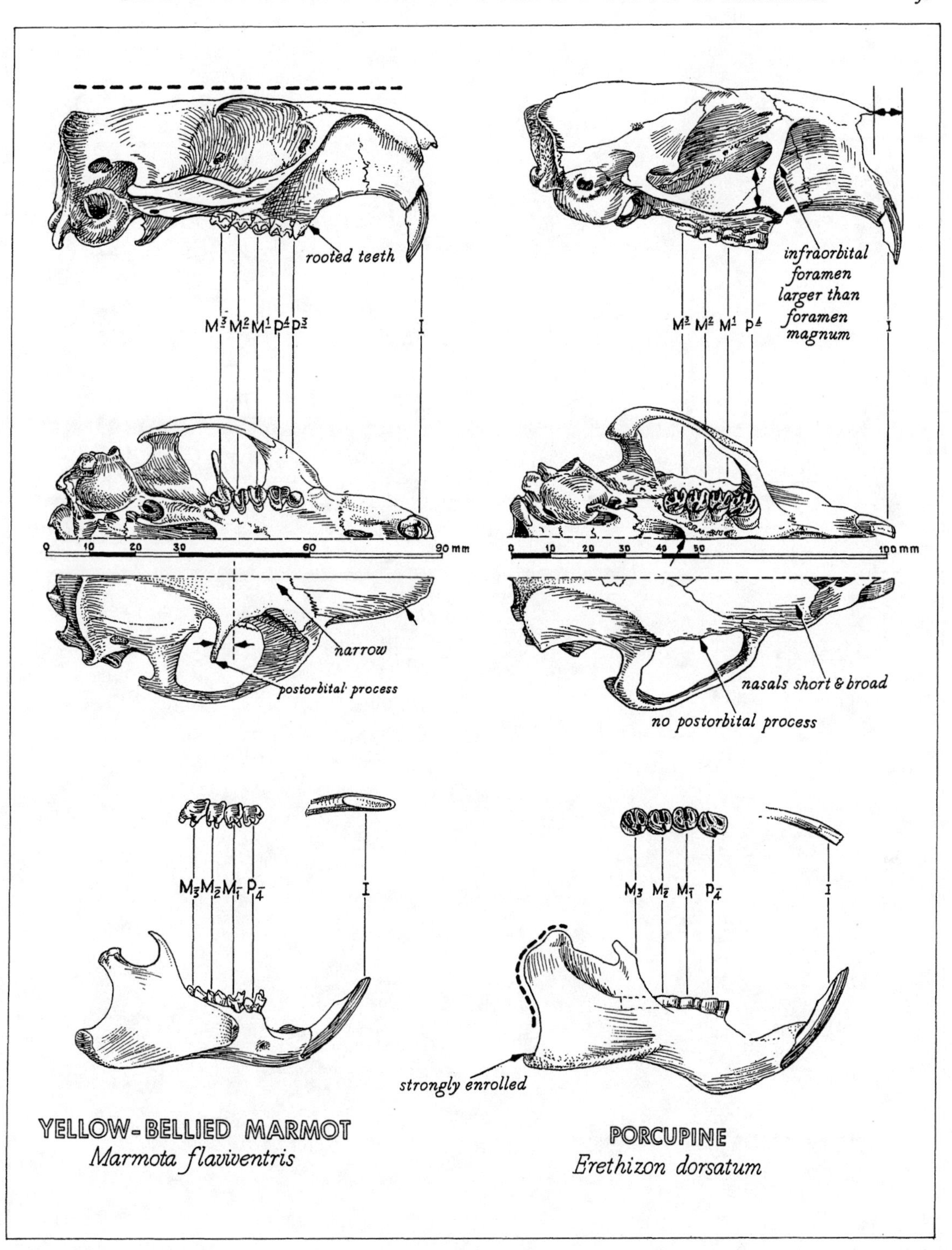
rooted teeth
M3 M2 M1 P4 P3
I
infraorbital foramen larger than foramen magnum
M3 M2 M1 P4
I
0 10 20 30 60 90 mm
0 10 20 30 40 50 100 mm
narrow
postorbital process
nasals short & broad
no postorbital process
M3 M2 M1 P4
I
M3 M2 M1 P4
I
strongly enrolled
YELLOW-BELLIED MARMOT
Marmota flaviventris
PORCUPINE
Erethizon dorsatum

FIGURE 8

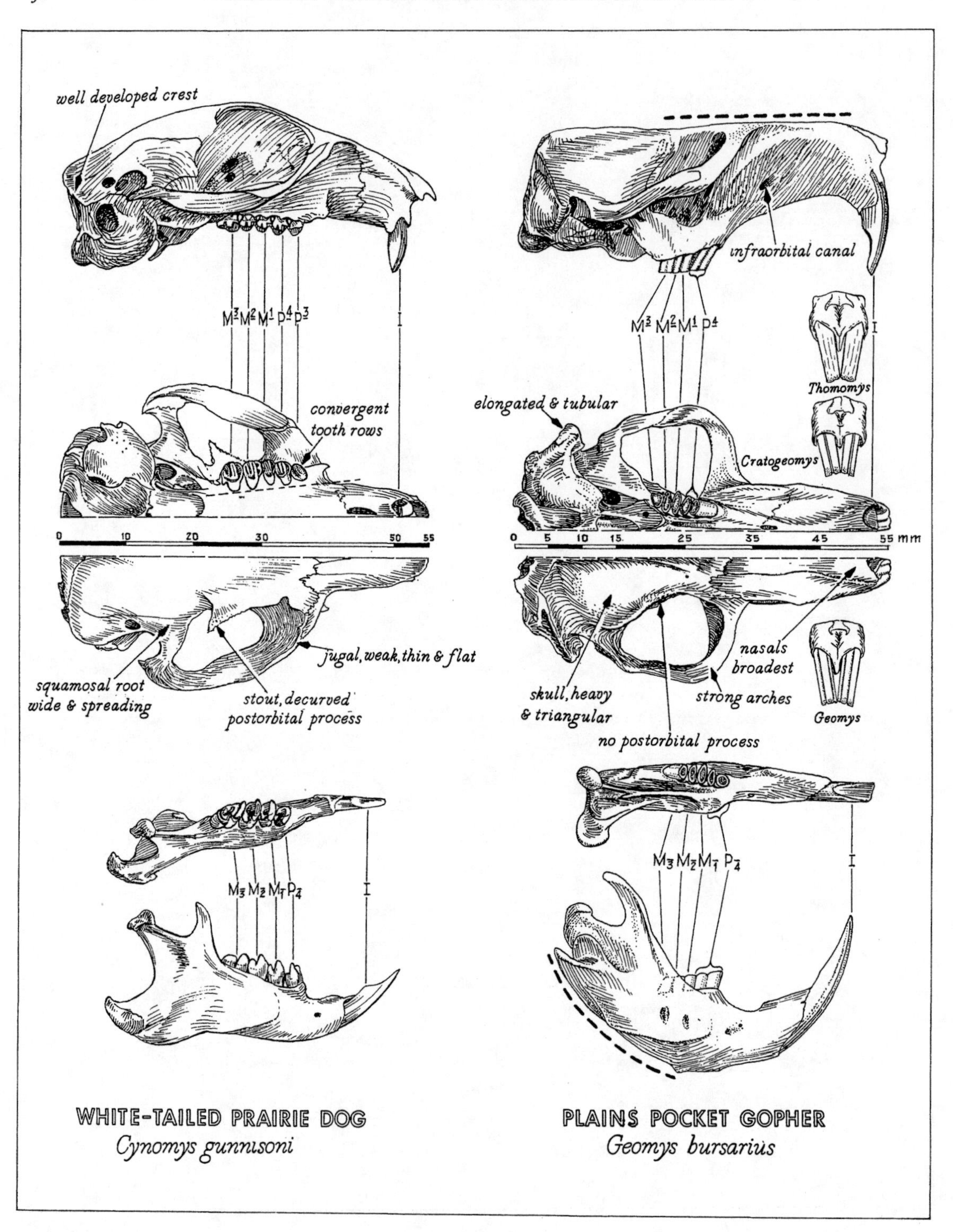

well developed crest
M3 M2 M1 P4 P3
I
convergent tooth rows
squamosal root wide & spreading
stout, decurved postorbital process
jugal, weak, thin & flat
0 10 20 30 50 55
M3 M2 M1 P4
I
WHITE-TAILED PRAIRIE DOG
Cynomys gunnisoni
infraorbital canal
M3 M2 M1 P4
I
Thomomys
Cratogeomys
elongated & tubular
0 5 10 15 25 35 45 55 mm
nasals broadest
skull, heavy & triangular
strong arches
Geomys
no postorbital process
M3 M2 M1 P4
I
PLAINS POCKET GOPHER
Geomys bursarius

FIGURE 9

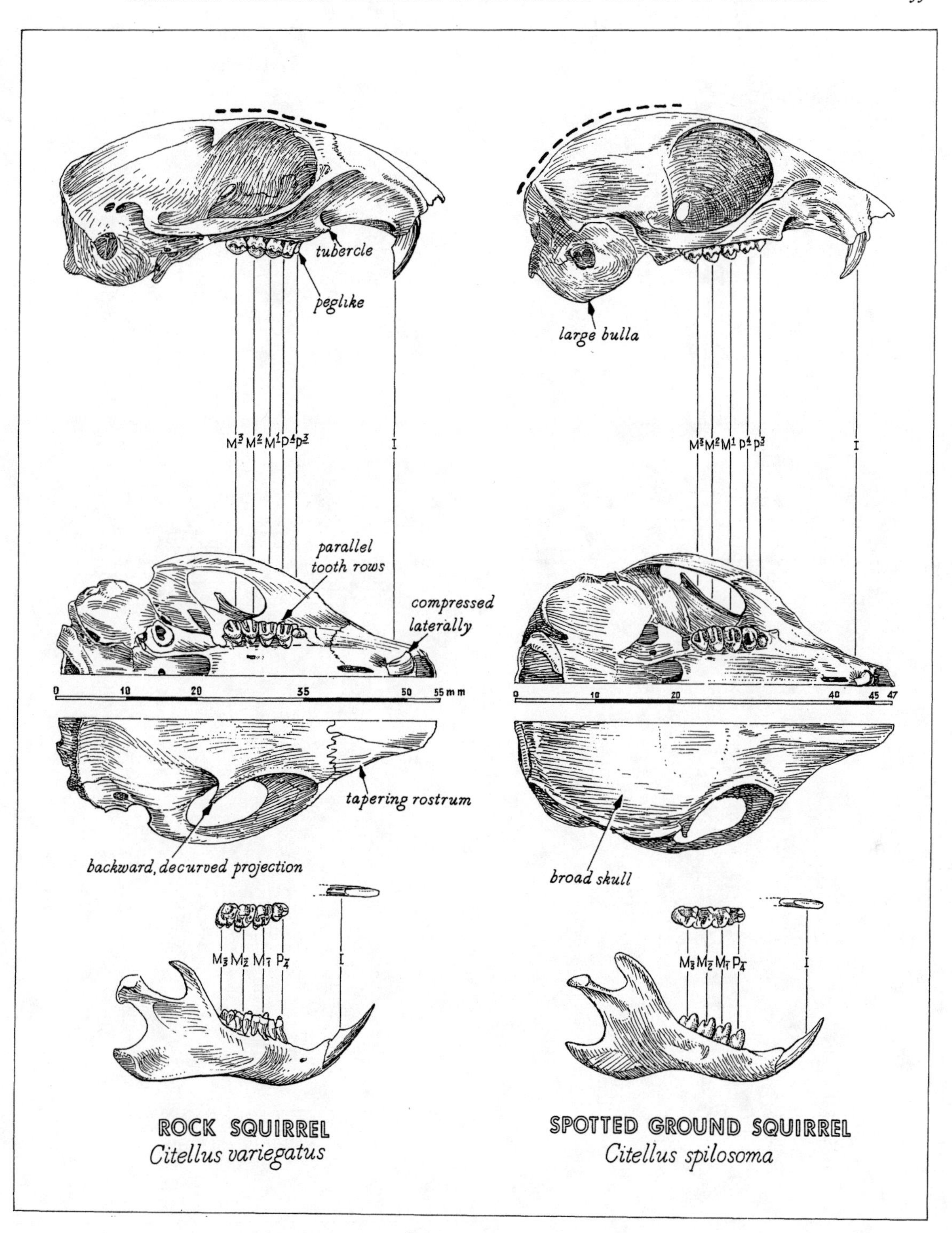
tubercle
peglike
large bulla
parallel tooth rows
compressed laterally
0 10 20 35 50 55 mm
0 10 20 40 45 47
tapering rostrum
backward, decurved projection
broad skull
ROCK SQUIRREL
Citellus variegatus
SPOTTED GROUND SQUIRREL
Citellus spilosoma

FIGURE 10

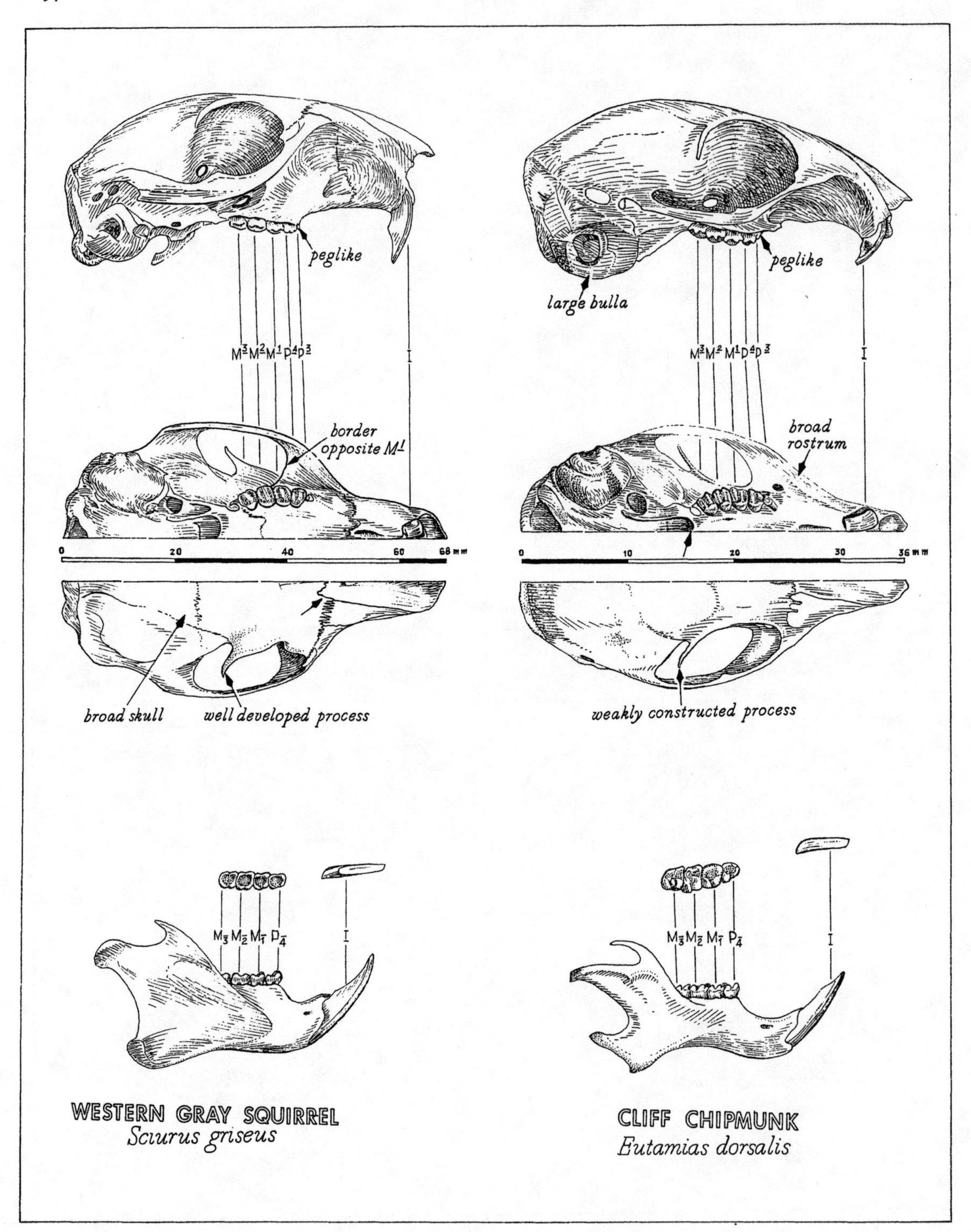
peglike
peglike
large bulla
M3 M2 M1 P4 P3
I
M3 M2 M1 P4 P3
I
border opposite M1
broad rostrum
0 20 40 60 68 mm
0 10 20 30 36 mm
broad skull
well developed process
weakly constructed process
M3 M2 M1 P4
I
M3 M2 M1 P4
I
WESTERN GRAY SQUIRREL
Sciurus griseus
CLIFF CHIPMUNK
Eutamias dorsalis

FIGURE 11

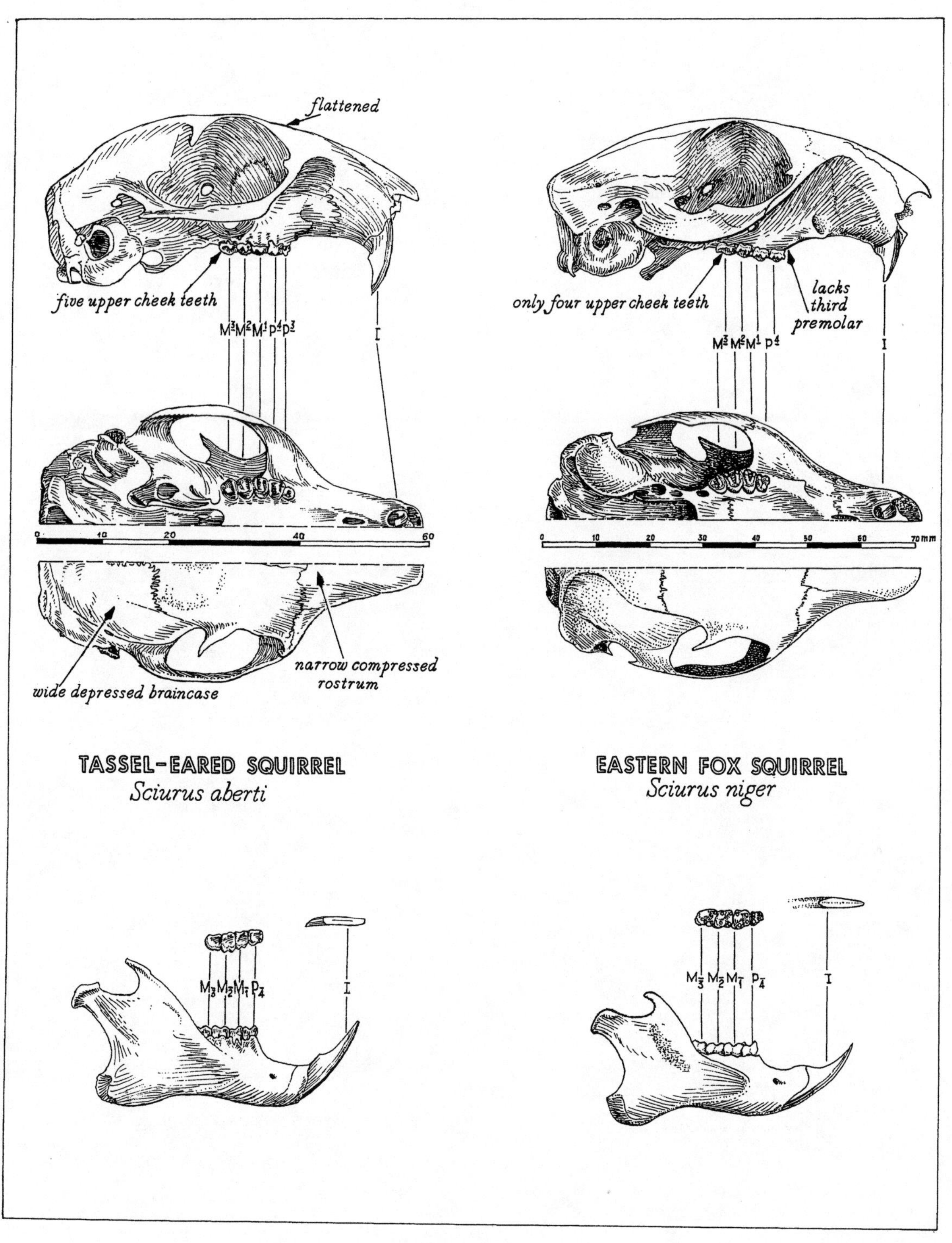
flattened
five upper cheek teeth
M3 M2 M1 P4 P3
I
only four upper cheek teeth
lacks third premolar
M3 M2 M1 P4
I
0 10 20 40 60
0 10 20 30 40 50 60 70 mm
narrow compressed rostrum
wide depressed braincase
TASSEL-EARED SQUIRREL
Sciurus aberti
EASTERN FOX SQUIRREL
Sciurus niger
M3 M2 M1 P4
I
M3 M2 M1 P4
I

FIGURE 12

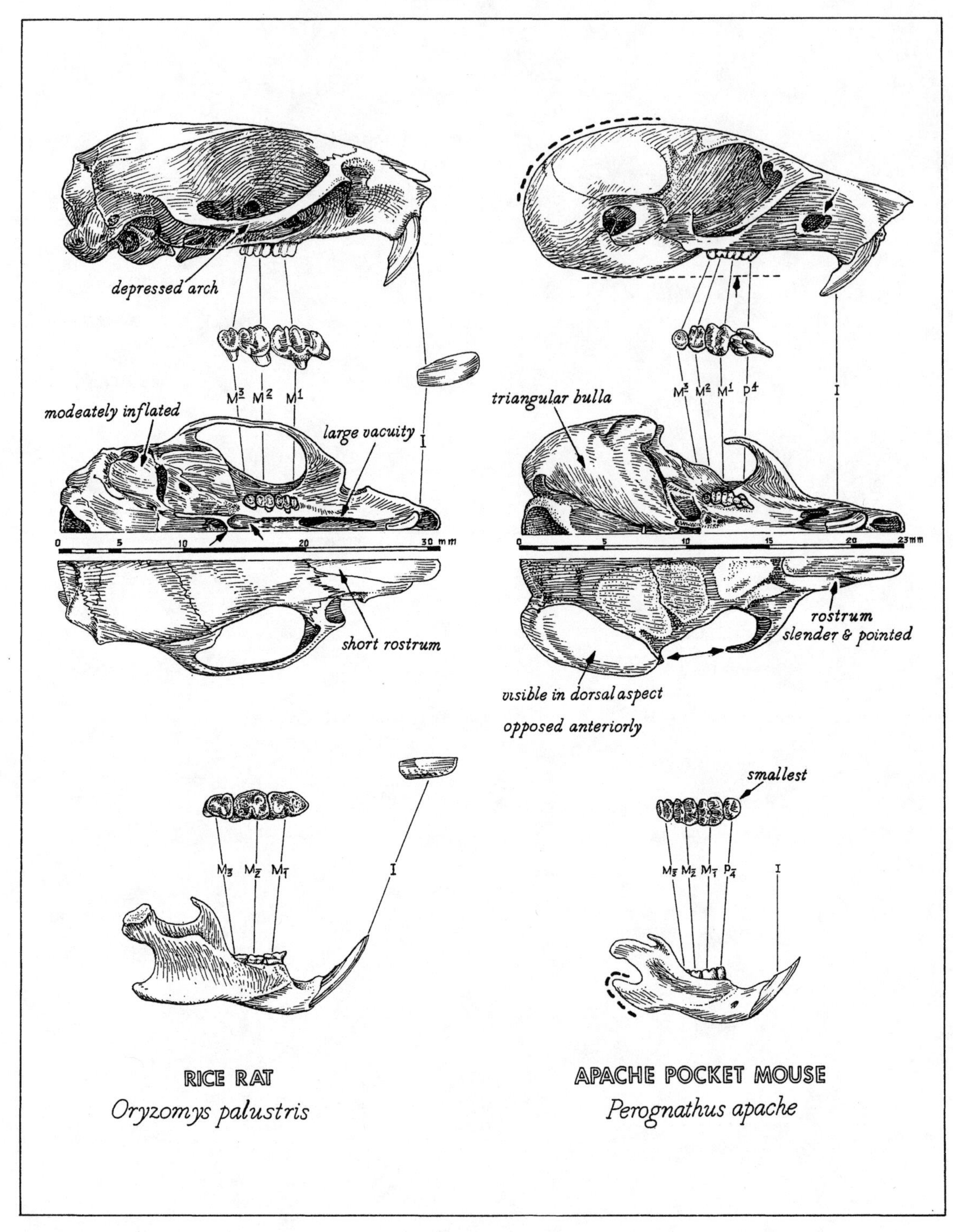
depressed arch
M3 M2 M1
modeately inflated
large vacuity
I
0 5 10 20 30 mm
short rostrum
M3 M2 M1
I
RICE RAT
Oryzomys palustris
triangular bulla
M3 M2 M1 P4
I
0 5 10 15 20 23mm
rostrum
slender & pointed
visible in dorsal aspect
opposed anteriorly
smallest
M3 M2 M1 P4
I
APACHE POCKET MOUSE
Perognathus apache

FIGURE 13

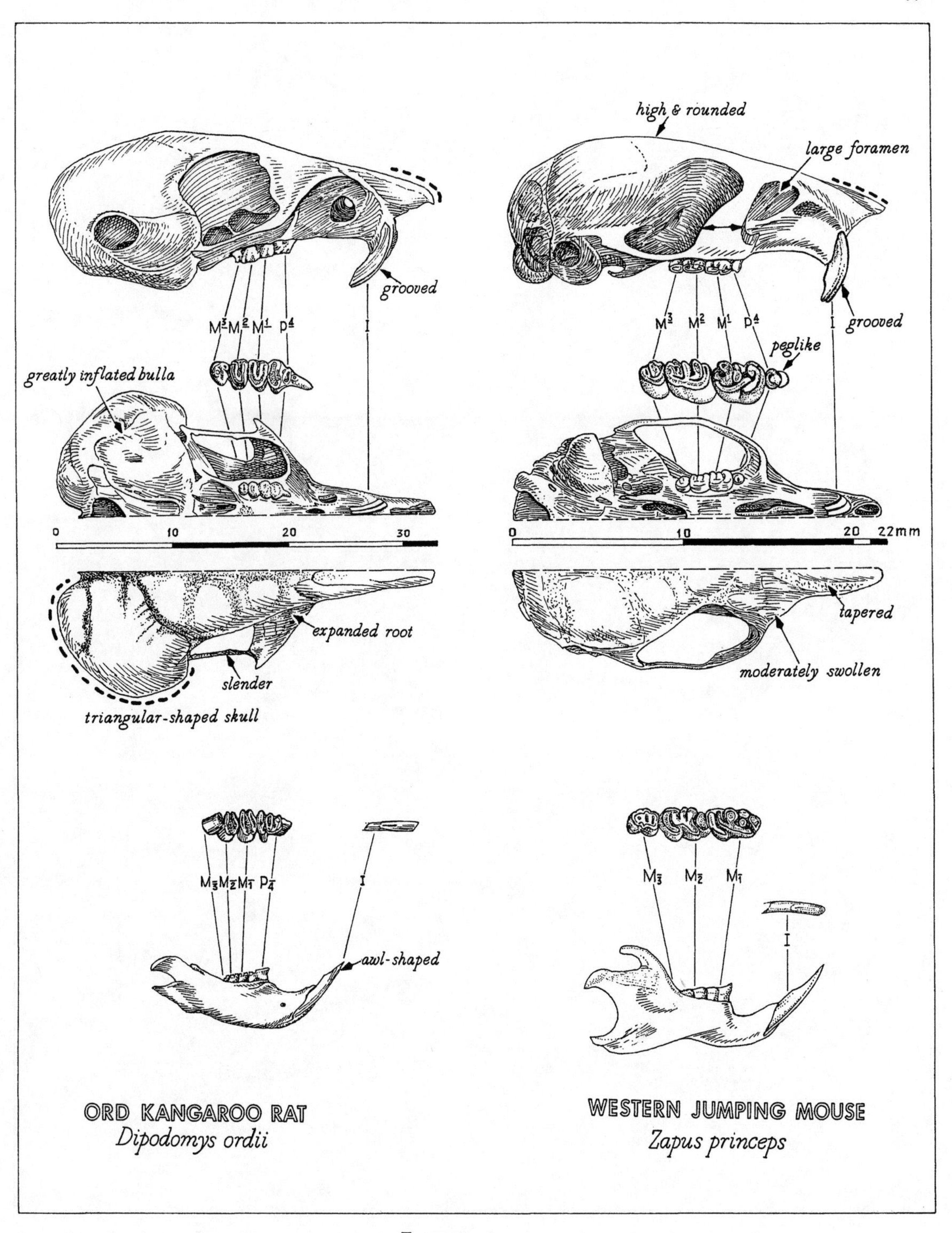

grooved
M3 M2 M1 P4
I
greatly inflated bulla
0 10 20 30
expanded root
slender
triangular-shaped skull
M3 M2 M1 P4
I
awl-shaped
ORD KANGAROO RAT
Dipodomys ordii
high & rounded
large foramen
M3 M2 M1 P4
I grooved
peglike
0 10 20 22mm
tapered
moderately swollen
M3 M2 M1
I
WESTERN JUMPING MOUSE
Zapus princeps

FIGURE 14

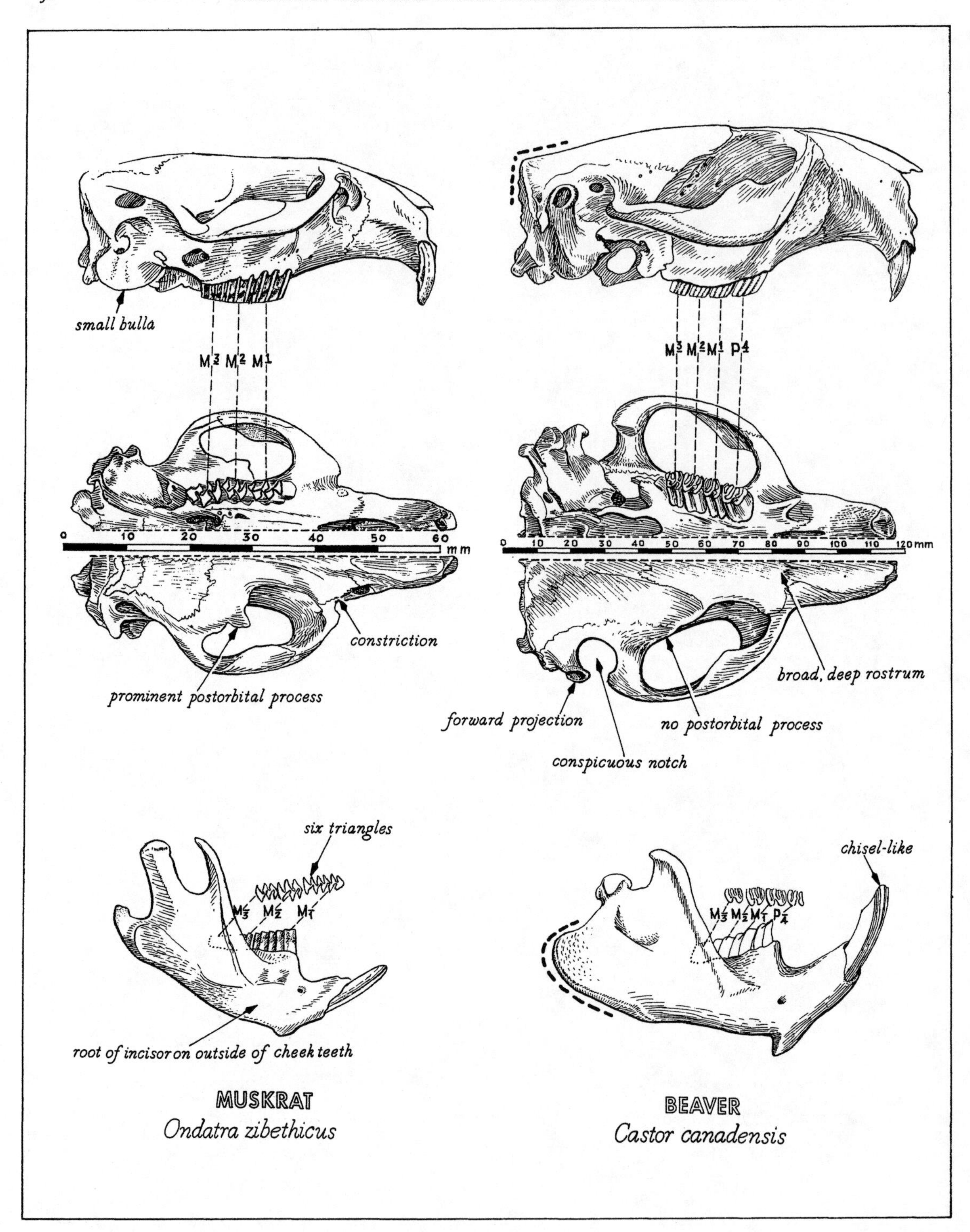
small bulla
M3 M2 M1
M3 M2 M1 P4
constriction
prominent postorbital process
broad, deep rostrum
forward projection
no postorbital process
conspicuous notch
six triangles
chisel-like
root of incisor on outside of cheek teeth
MUSKRAT
Ondatra zibethicus
BEAVER
Castor canadensis

FIGURE 15

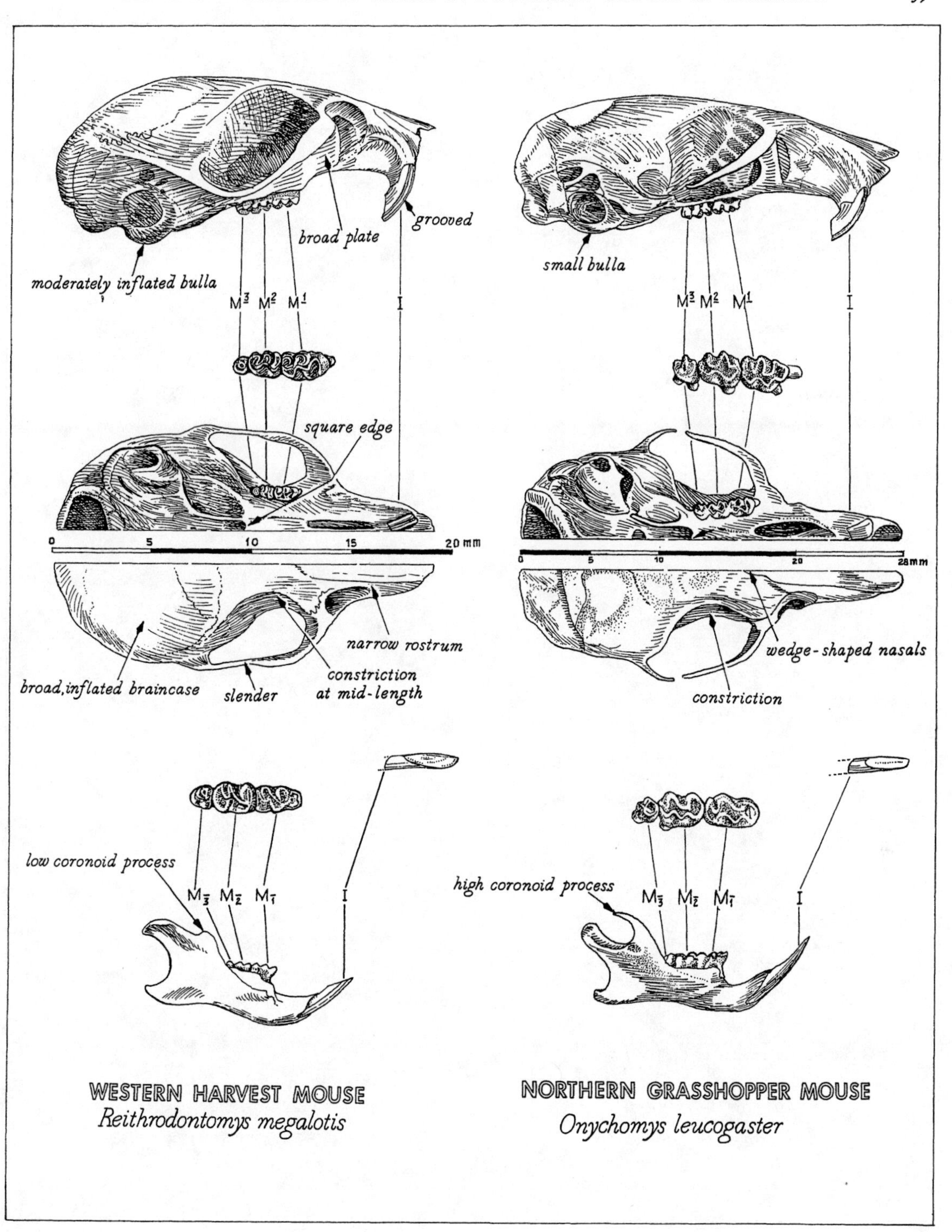
grooved
broad plate
moderately inflated bulla
M^3 M^2 M^1 I
square edge
0 5 10 15 20 mm
narrow rostrum
constriction
at mid-length
broad, inflated braincase
slender
small bulla
M^3 M^2 M^1 I
0 5 10 20 28 mm
wedge-shaped nasals
constriction
low coronoid process
M_3 M_2 M_1 I
high coronoid process
M_3 M_2 M_1 I
WESTERN HARVEST MOUSE
Reithrodontomys megalotis
NORTHERN GRASSHOPPER MOUSE
Onychomys leucogaster

FIGURE 16

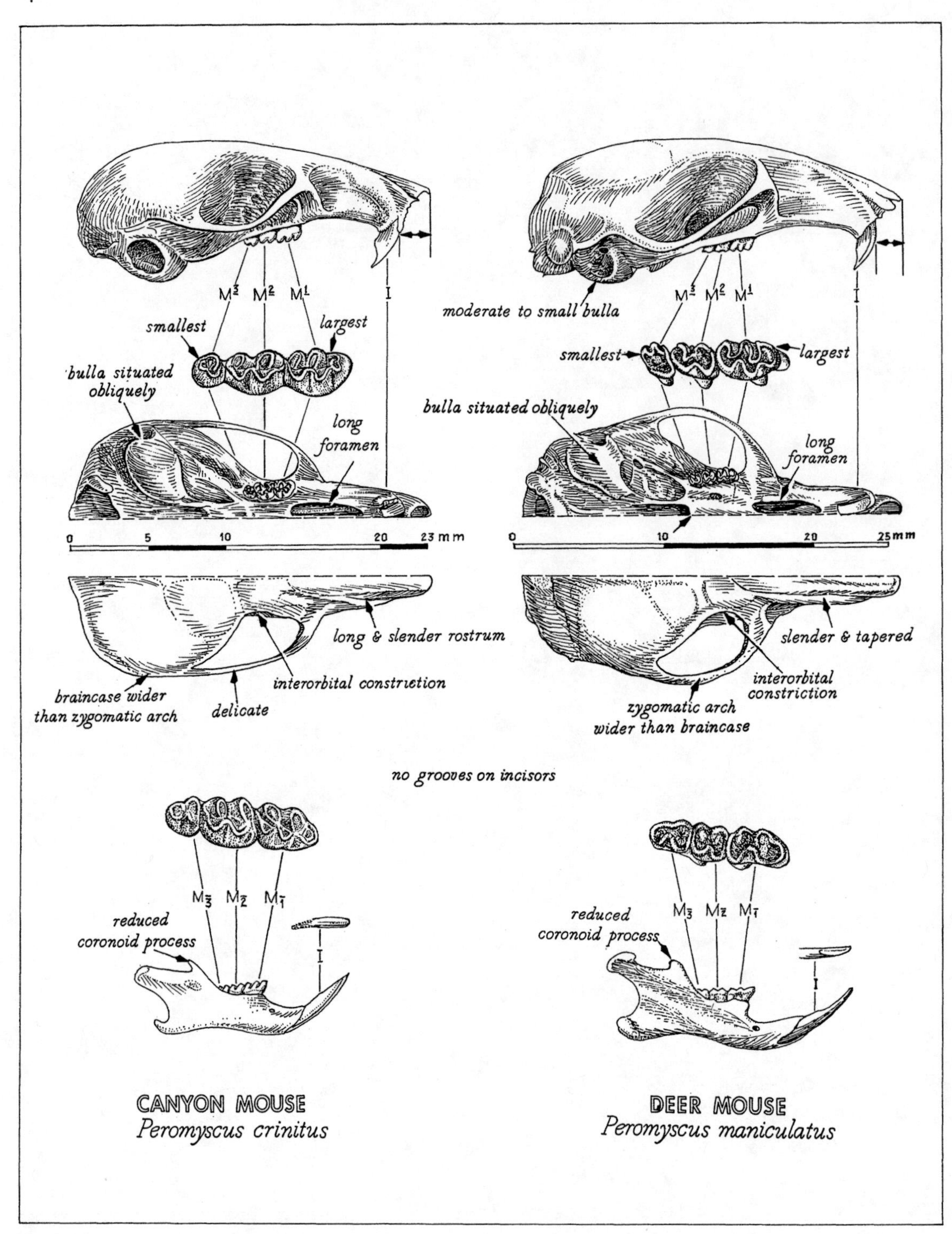
M3 M2 M1
I
smallest
largest
bulla situated obliquely
long foramen
0 5 10 20 23 mm
long & slender rostrum
interorbital constriction
braincase wider than zygomatic arch
delicate
moderate to small bulla
M3 M2 M1
I
smallest
largest
bulla situated obliquely
long foramen
0 10 20 25 mm
slender & tapered
interorbital constriction
zygomatic arch wider than braincase
no grooves on incisors
M3 M2 M1
reduced coronoid process
I
M3 M2 M1
reduced coronoid process
I
CANYON MOUSE
Peromyscus crinitus
DEER MOUSE
Peromyscus maniculatus

FIGURE 17

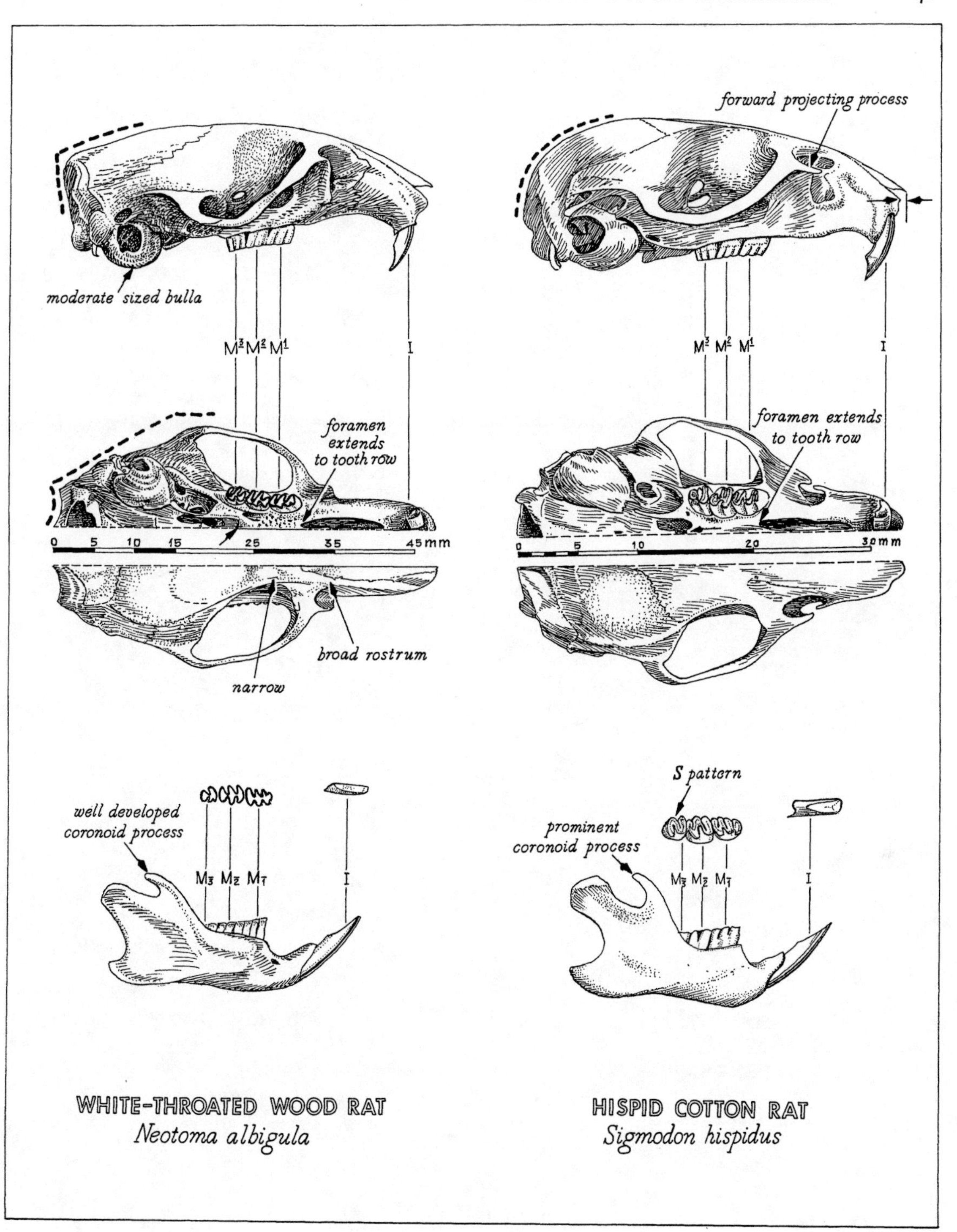

forward projecting process
moderate sized bulla
M3 M2 M1
I
foramen extends to tooth row
foramen extends to tooth row
0 5 10 15 25 35 45mm
0 5 10 20 30mm
broad rostrum
narrow
S pattern
well developed coronoid process
prominent coronoid process
M3 M2 M1
I
WHITE-THROATED WOOD RAT
Neotoma albigula
HISPID COTTON RAT
Sigmodon hispidus

FIGURE 18

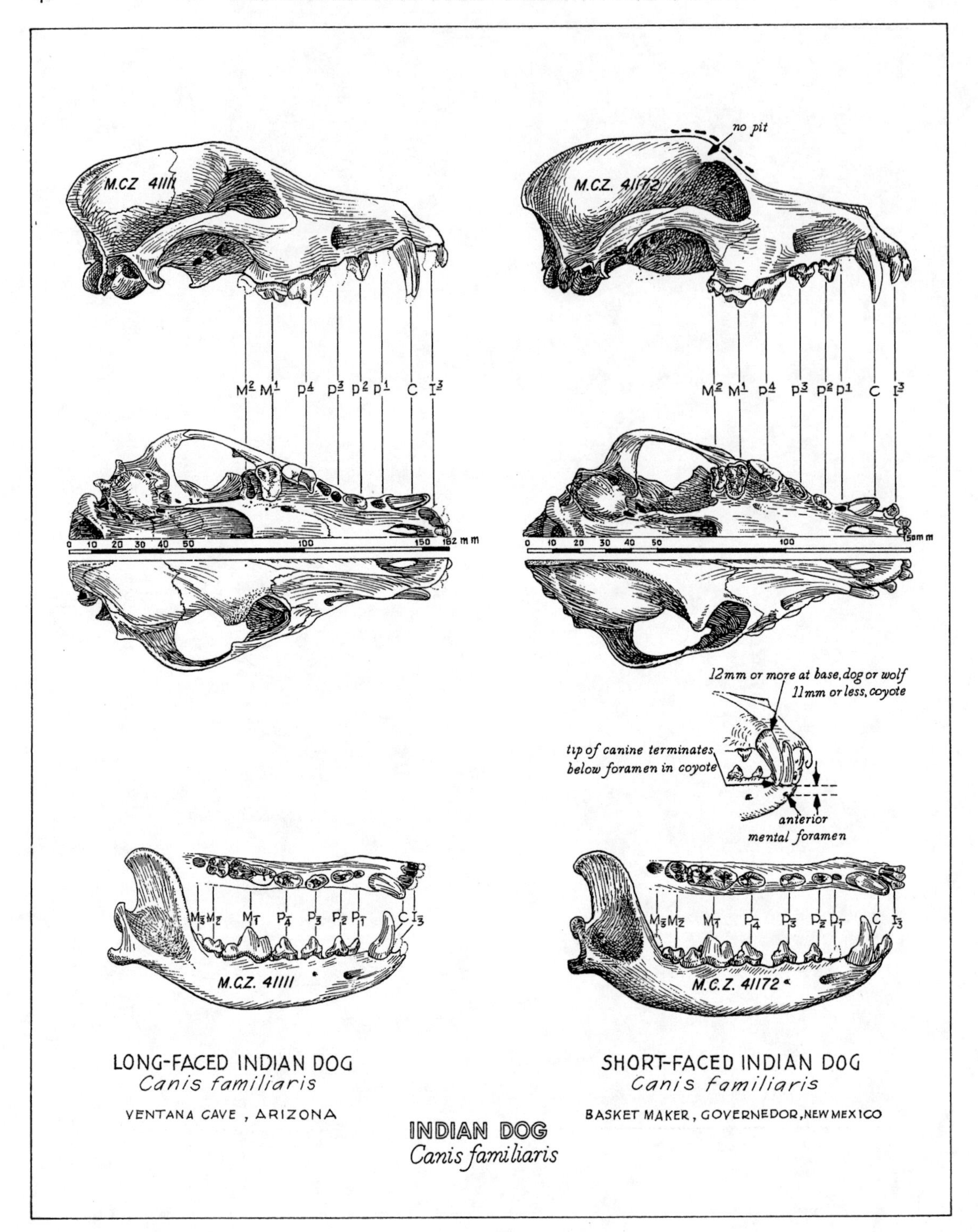
no pit
M.CZ 4111
M.C.Z. 41172
M2 M1 P4 P3 P2 P1 C I3
12mm or more at base, dog or wolf
11mm or less, coyote
tip of canine terminates below foramen in coyote
anterior mental foramen
M3 M2 M1 P4 P3 P2 P1 C I3
M.C.Z. 41111
M.C.Z. 41172
LONG-FACED INDIAN DOG
Canis familiaris
VENTANA CAVE, ARIZONA
SHORT-FACED INDIAN DOG
Canis familiaris
BASKET MAKER, GOVERNEDOR, NEW MEXICO
INDIAN DOG
Canis familiaris

FIGURE 19

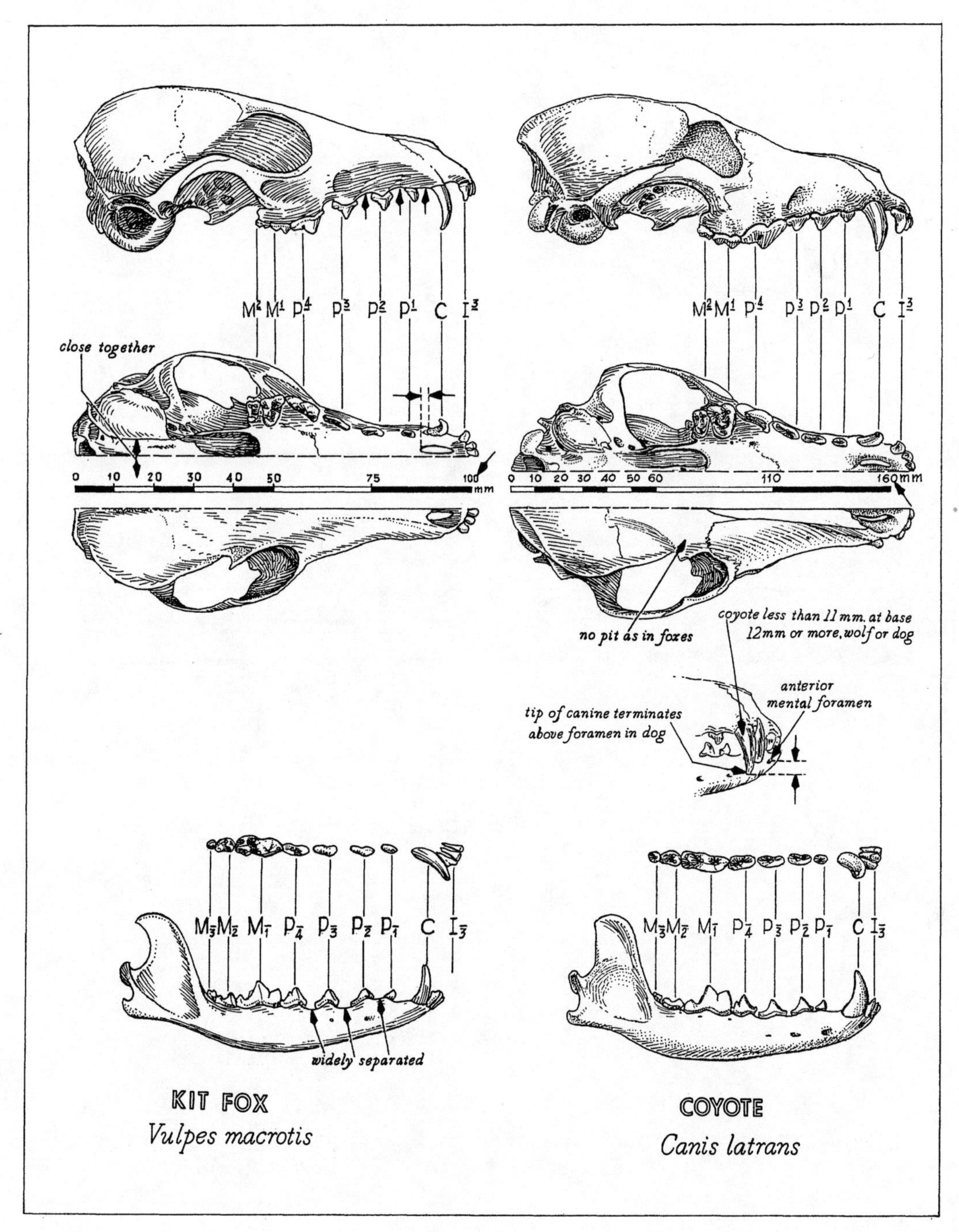

close together
no pit as in foxes
coyote less than 11 mm. at base
12 mm or more, wolf or dog
anterior
mental foramen
tip of canine terminates
above foramen in dog
widely separated
KIT FOX
Vulpes macrotis
COYOTE
Canis latrans

FIGURE 20

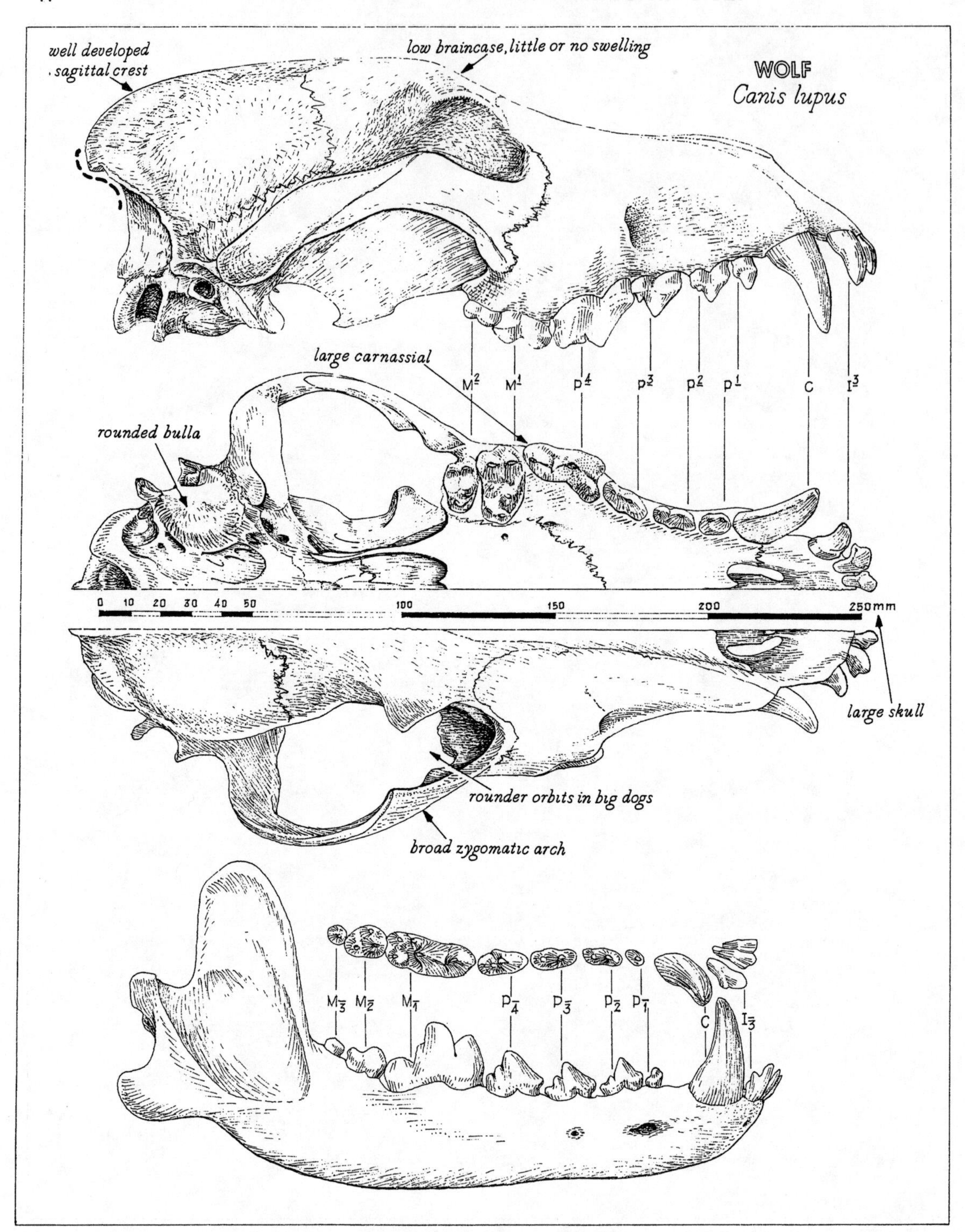
well developed sagittal crest
low braincase, little or no swelling
WOLF
Canis lupus
large carnassial
M2 M1 P4 P3 P2 P1 C I3
rounded bulla
0 10 20 30 40 50 100 150 200 250mm
large skull
rounder orbits in big dogs
broad zygomatic arch
M3 M2 M1 P4 P3 P2 P1 C I3

FIGURE 21

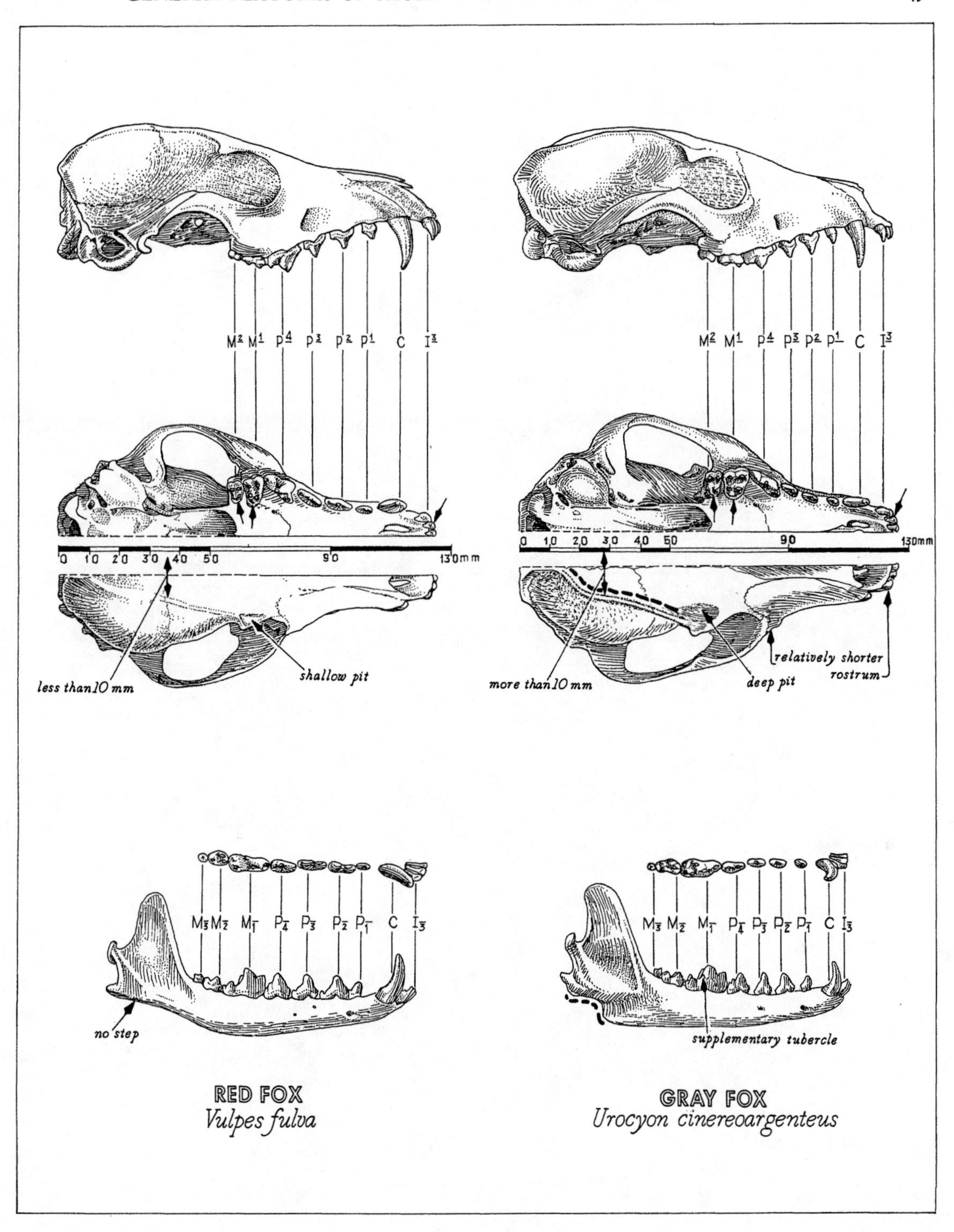

M2 M1 P4 P3 P2 P1 C I3
M2 M1 P4 P3 P2 P1 C I3
0 10 20 30 40 50 90 130mm
0 10 20 30 40 50 90 130mm
less than 10 mm
shallow pit
more than 10 mm
deep pit
relatively shorter rostrum
M3 M2 M1 P4 P3 P2 P1 C I3
M3 M2 M1 P4 P3 P2 P1 C I3
no step
supplementary tubercle
RED FOX
Vulpes fulva
GRAY FOX
Urocyon cinereoargenteus

FIGURE 22

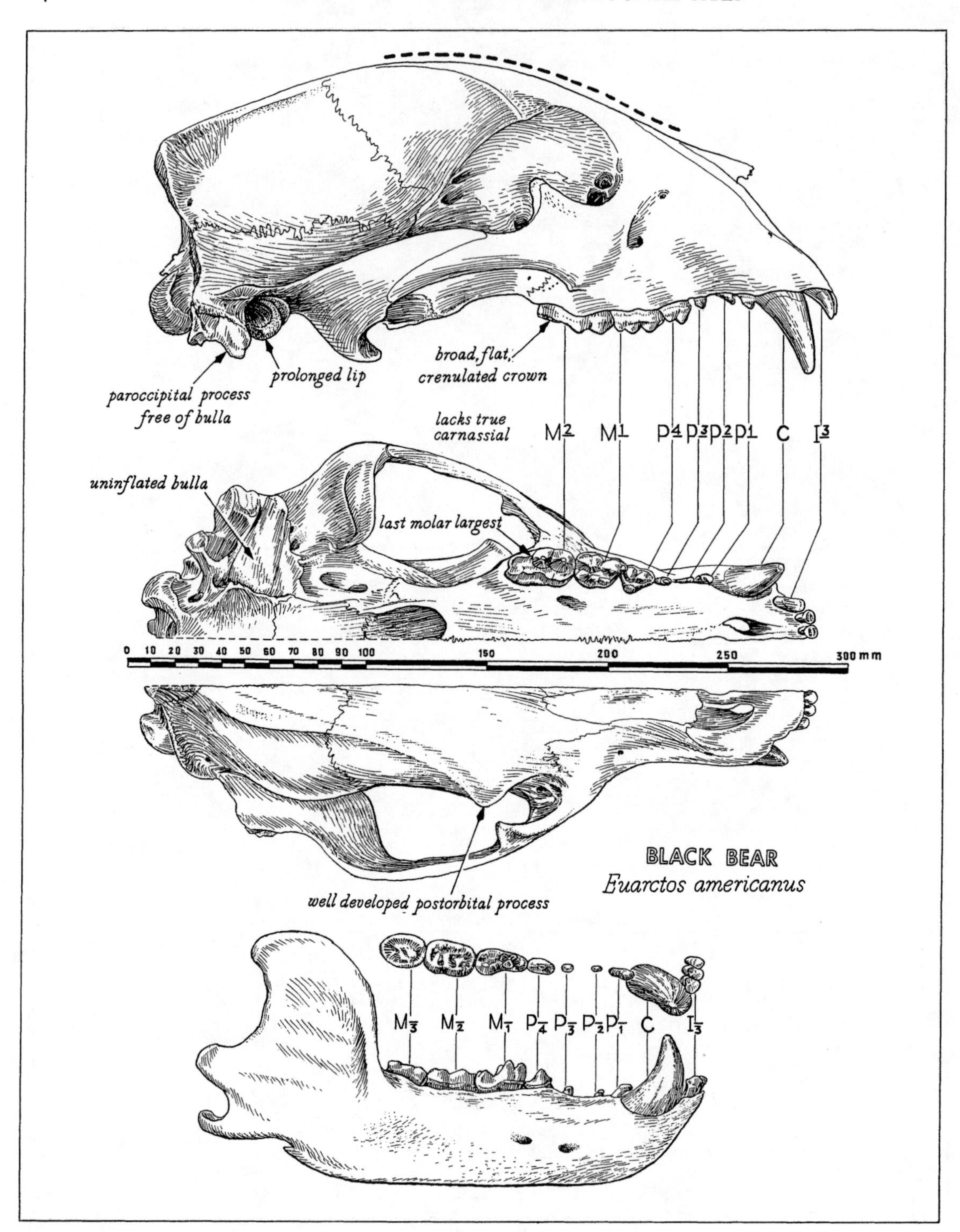
broad, flat, crenulated crown
prolonged lip
paroccipital process free of bulla
lacks true carnassial
M2 M1 P4 P3 P2 P1 C I3
uninflated bulla
last molar largest
0 10 20 30 40 50 60 70 80 90 100 150 200 250 300 mm
BLACK BEAR
Euarctos americanus
well developed postorbital process
M3 M2 M1 P4 P3 P2 P1 C I3

FIGURE 23

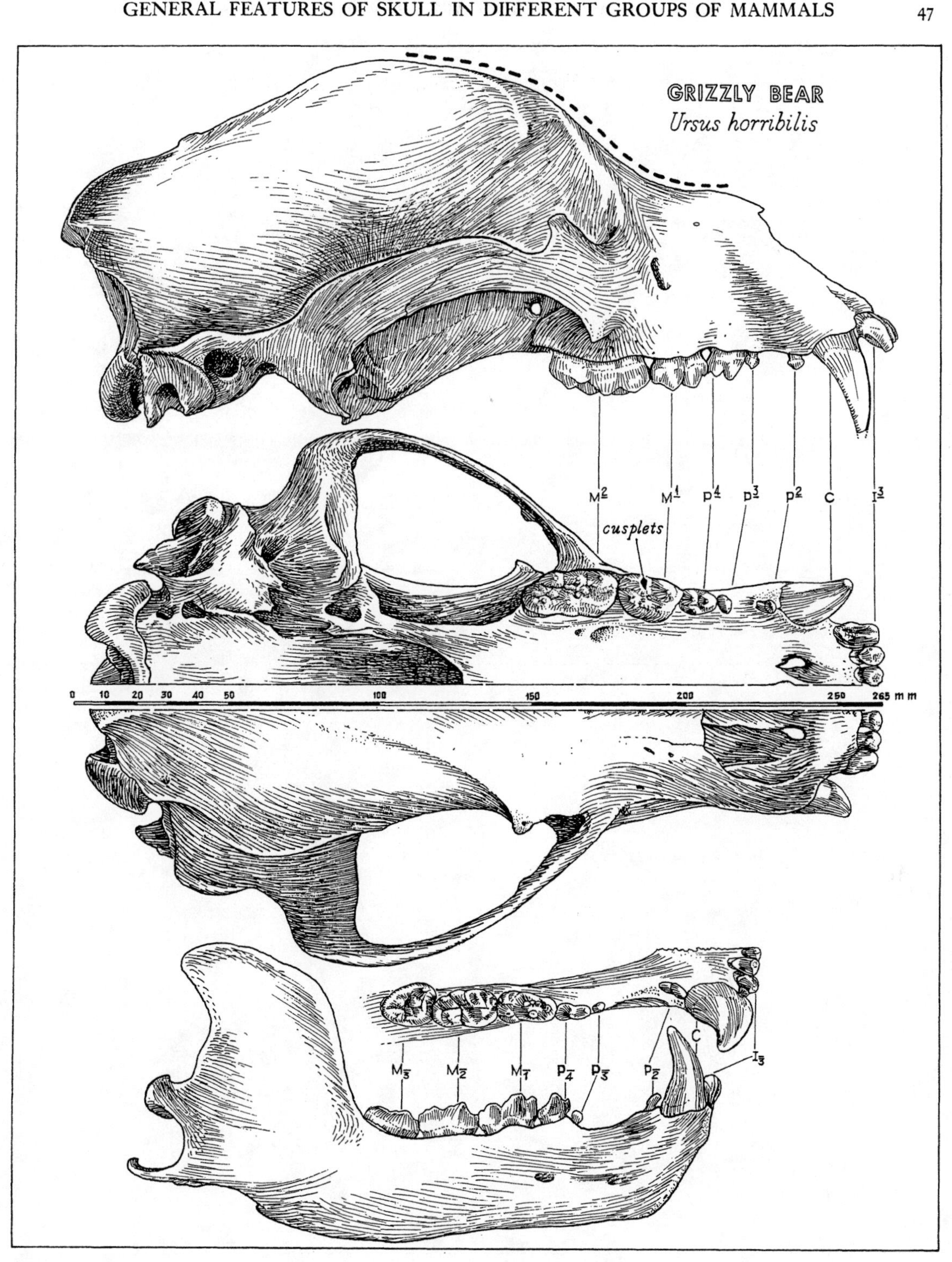
GRIZZLY BEAR
Ursus horribilis
M2
M1
P4
P3
P2
C
I3
cusplets
0 10 20 30 40 50 100 150 200 250 265 mm
M3
M2
M1
P4
P3
P2
C
I3

FIGURE 24

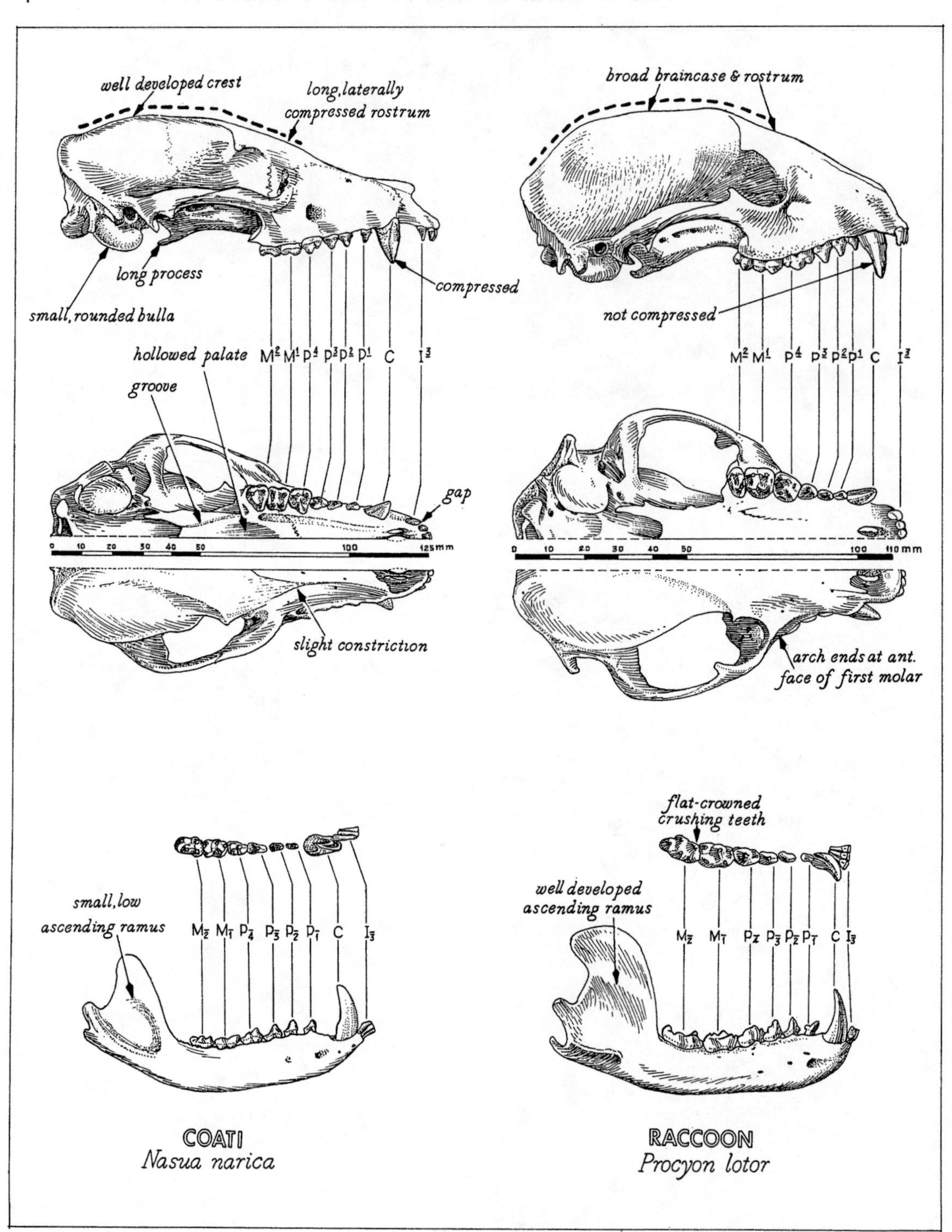

well developed crest
long, laterally compressed rostrum
long process
small, rounded bulla
compressed
broad braincase & rostrum
not compressed
hollowed palate
groove
gap
slight constriction
arch ends at ant. face of first molar
flat-crowned crushing teeth
small, low ascending ramus
well developed ascending ramus
COATI
Nasua narica
RACCOON
Procyon lotor

FIGURE 25

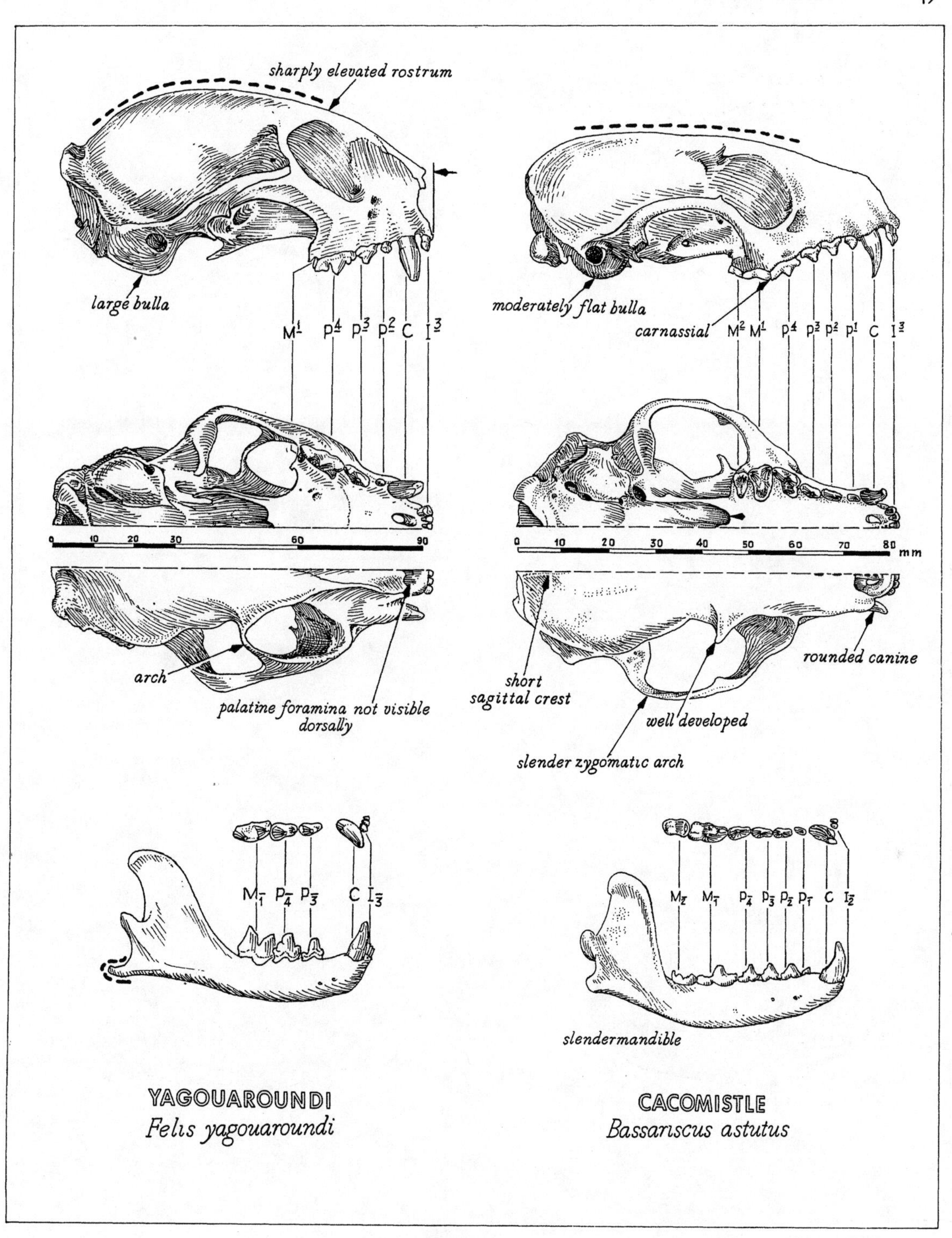
sharply elevated rostrum
large bulla
M1 P4 P3 P2 C I3
moderately flat bulla
carnassial
M2 M1 P4 P3 P2 P1 C I3
0 10 20 30 60 90
0 10 20 30 40 50 60 70 80 mm
arch
palatine foramina not visible dorsally
short sagittal crest
well developed
slender zygomatic arch
rounded canine
M1 P4 P3 C I3
M2 M1 P4 P3 P2 P1 C I2
slendermandible
YAGOUAROUNDI
Felis yagouaroundi
CACOMISTLE
Bassariscus astutus

FIGURE 26

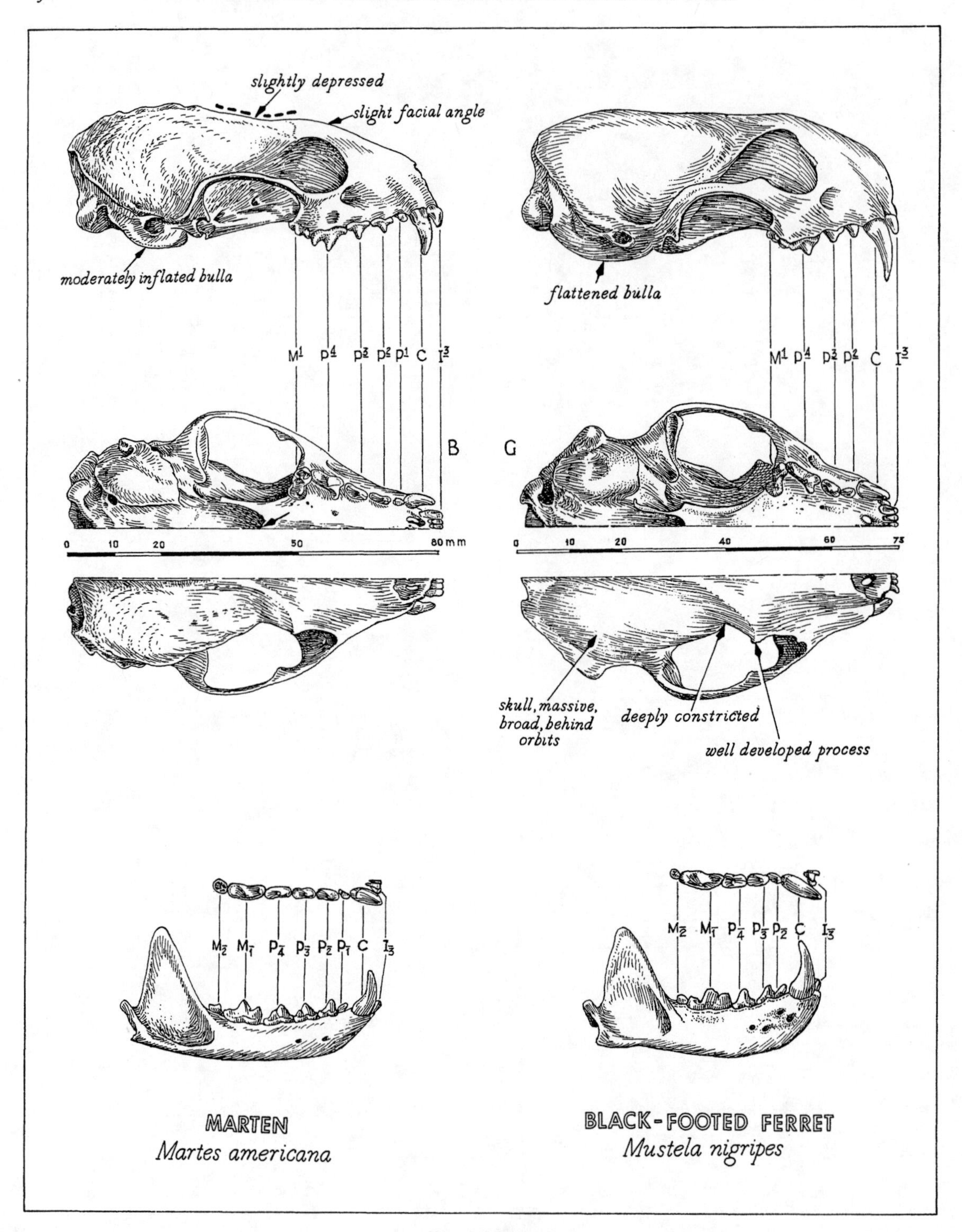
slightly depressed
slight facial angle
moderately inflated bulla
flattened bulla
M1 P4 P3 P2 P1 C I3
M1 P4 P3 P2 C I3
B
G
0 10 20 50 80 mm
0 10 20 40 60 73
skull, massive, broad, behind orbits
deeply constricted
well developed process
M2 M1 P4 P3 P2 P1 C I3
M2 M1 P4 P3 P2 C I3
MARTEN
Martes americana
BLACK-FOOTED FERRET
Mustela nigripes

FIGURE 27

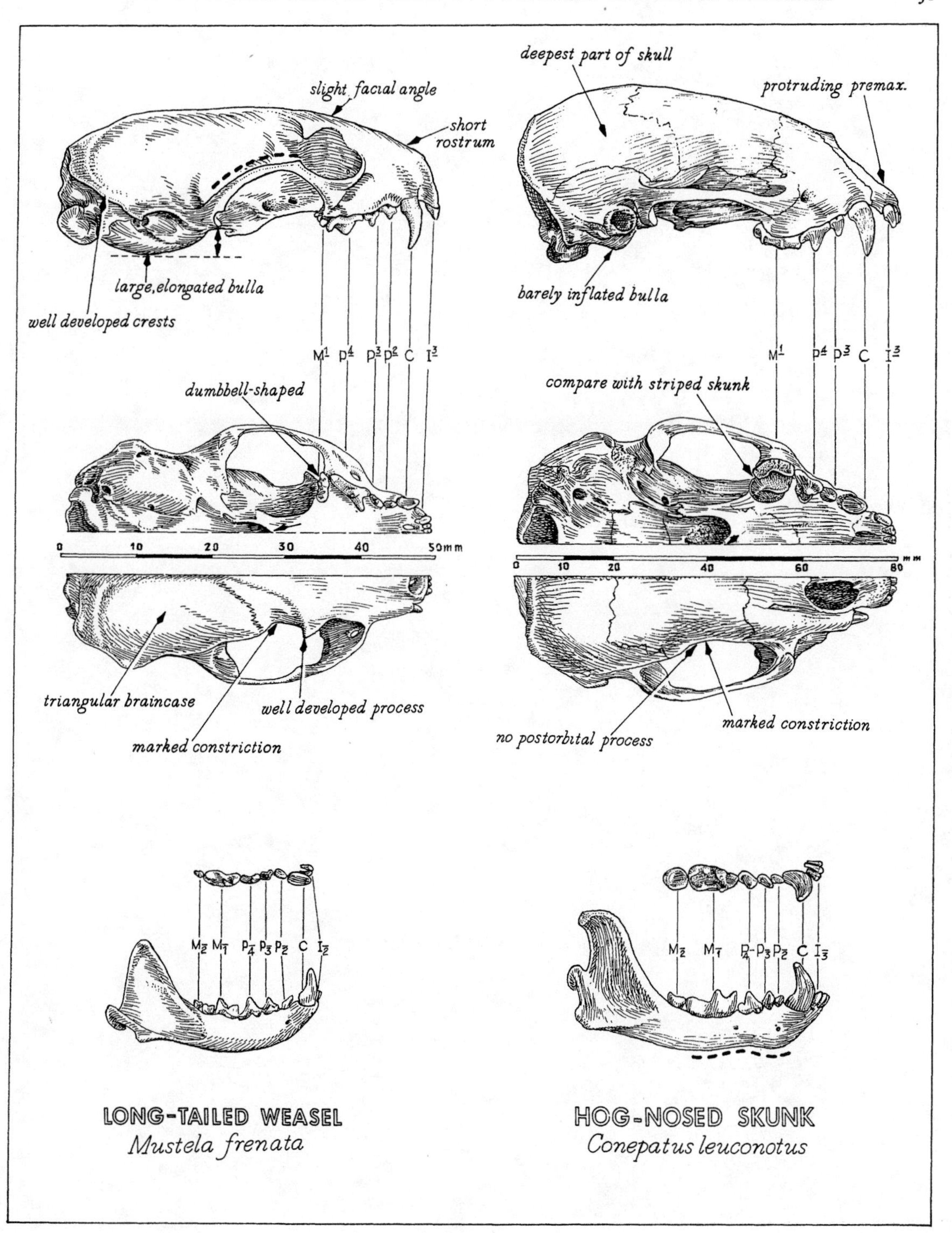
slight facial angle
short rostrum
large, elongated bulla
well developed crests
deepest part of skull
protruding premax.
barely inflated bulla
dumbbell-shaped
compare with striped skunk
triangular braincase
well developed process
marked constriction
no postorbital process
marked constriction
LONG-TAILED WEASEL
Mustela frenata
HOG-NOSED SKUNK
Conepatus leuconotus

FIGURE 28

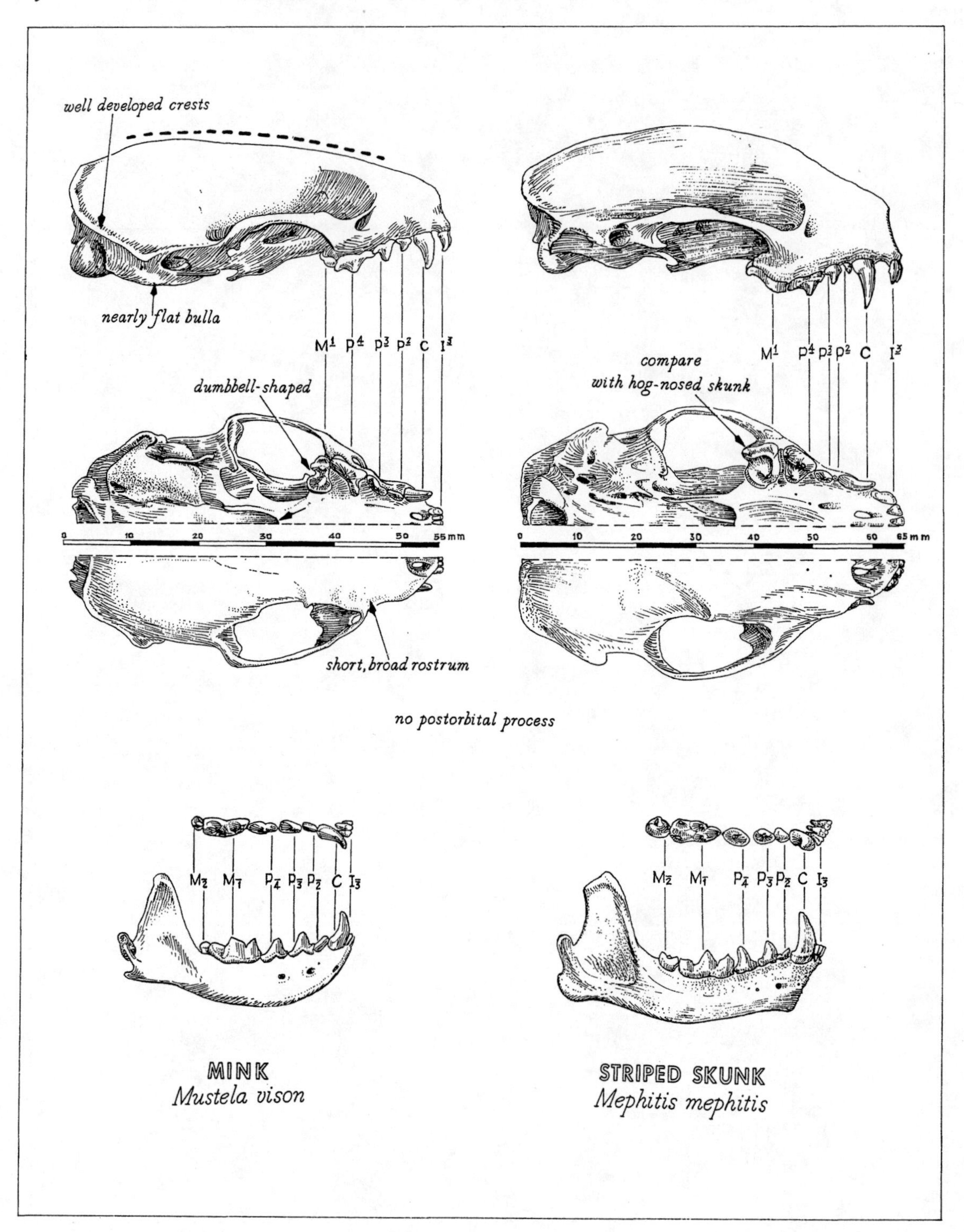

well developed crests
nearly flat bulla
M1 P4 P3 P2 C I3
dumbbell-shaped
compare
with hog-nosed skunk
M1 P4 P3 P2 C I3
0 10 20 30 40 50 55mm
0 10 20 30 40 50 60 65mm
short, broad rostrum
no postorbital process
M2 M1 P4 P3 P2 C I3
M2 M1 P4 P3 P2 C I3
MINK
Mustela vison
STRIPED SKUNK
Mephitis mephitis

FIGURE 29

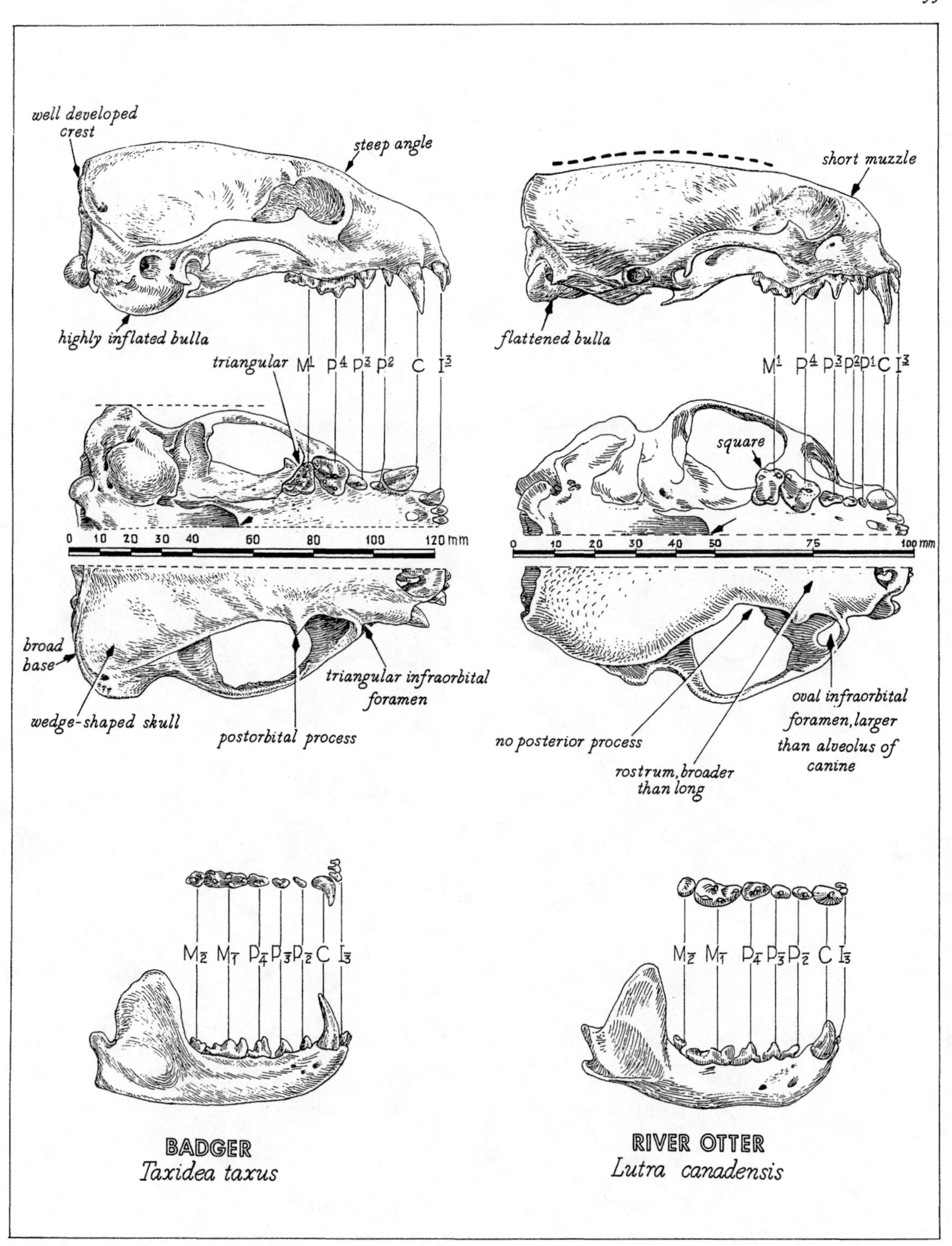

well developed crest
steep angle
highly inflated bulla
triangular
M1 P4 P3 P2 C I3
short muzzle
flattened bulla
M1 P4 P3 P2 P1 C I3
square
0 10 20 30 40 60 80 100 120 mm
0 10 20 30 40 50 75 100 mm
broad base
wedge-shaped skull
postorbital process
triangular infraorbital foramen
no posterior process
rostrum, broader than long
oval infraorbital foramen, larger than alveolus of canine
M2 M1 P4 P3 P2 C I3
M2 M1 P4 P3 P2 C I3
BADGER
Taxidea taxus
RIVER OTTER
Lutra canadensis

FIGURE 30

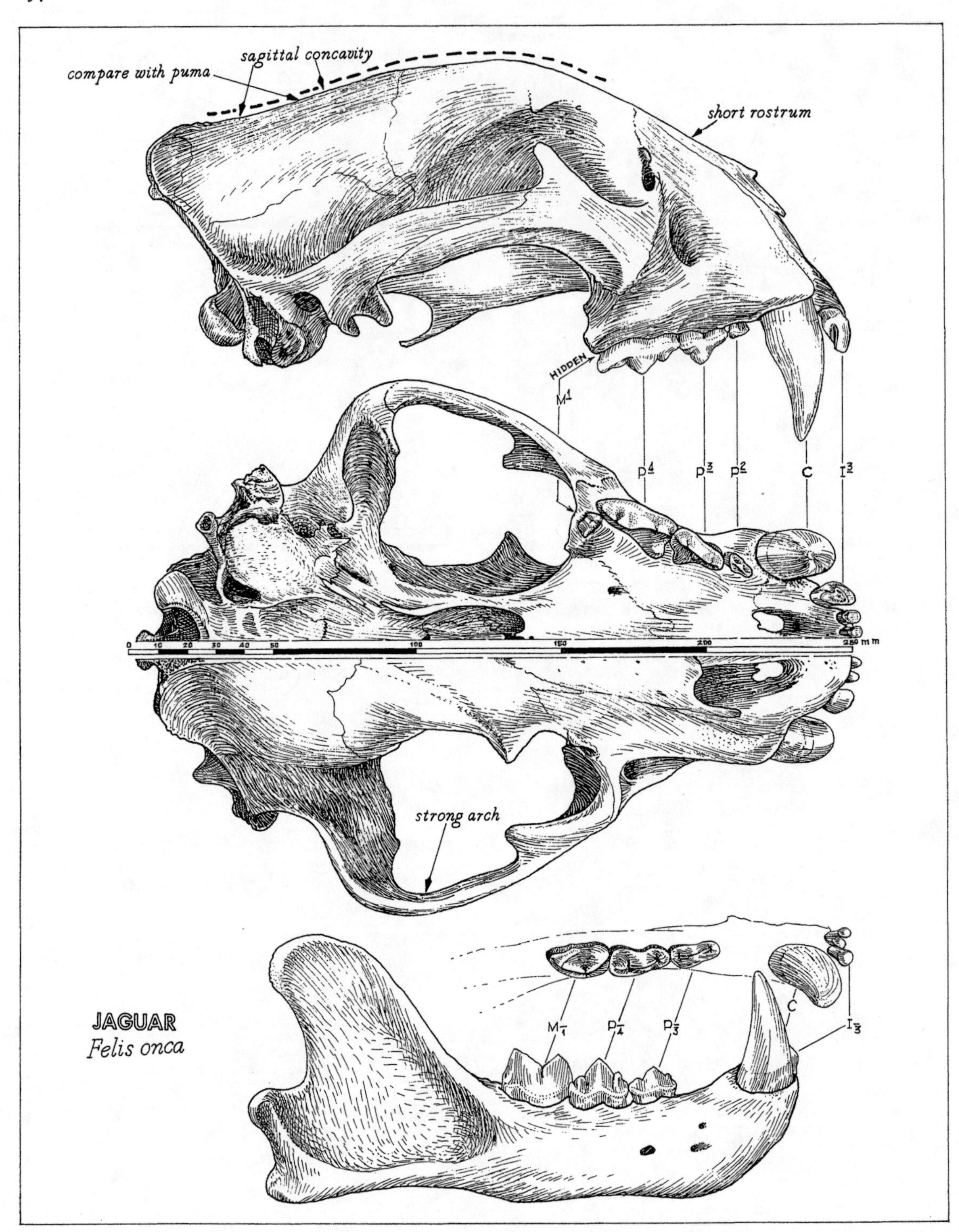

sagittal concavity
compare with puma
short rostrum
HIDDEN
M1
P4
P3
P2
C
I3
strong arch
JAGUAR
Felis onca
M1
P4
P3
C
I3

FIGURE 31

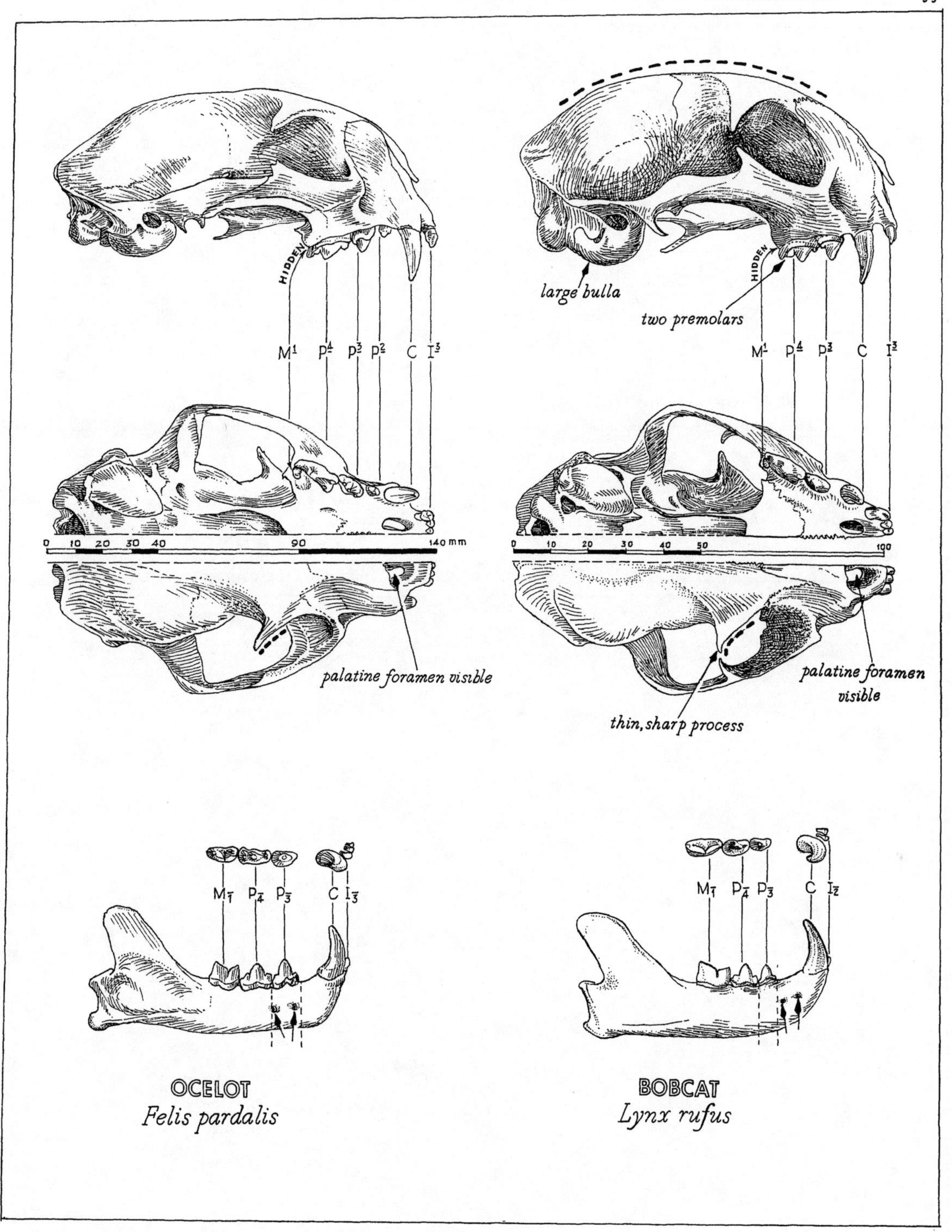
HIDDEN
M1 P4 P3 P2 C I3
palatine foramen visible
0 10 20 30 40 90 140 mm
large bulla
two premolars
HIDDEN
M1 P4 P3 C I3
0 10 20 30 40 50 100
palatine foramen visible
thin, sharp process
M1 P4 P3 C I3
M1 P4 P3 C I2
OCELOT
Felis pardalis
BOBCAT
Lynx rufus

FIGURE 32

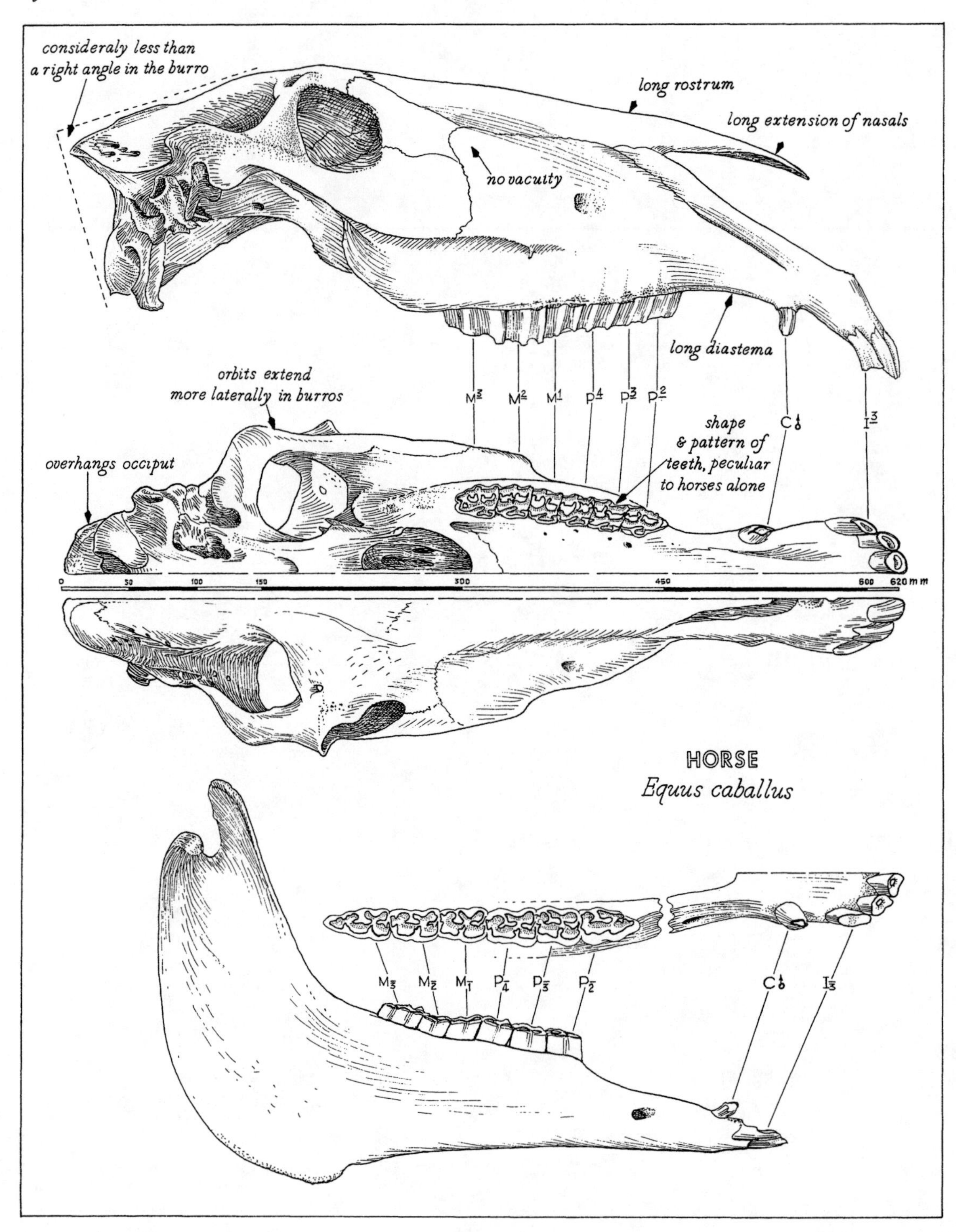
consideraly less than
a right angle in the burro
long rostrum
long extension of nasals
no vacuity
long diastema
orbits extend
more laterally in burros
overhangs occiput
shape
& pattern of
teeth, peculiar
to horses alone
M3 M2 M1 P4 P3 P2
C1 I3
0 50 100 150 300 450 600 620 mm
HORSE
Equus caballus
M3 M2 M1 P4 P3 P2
C1 I3

FIGURE 33

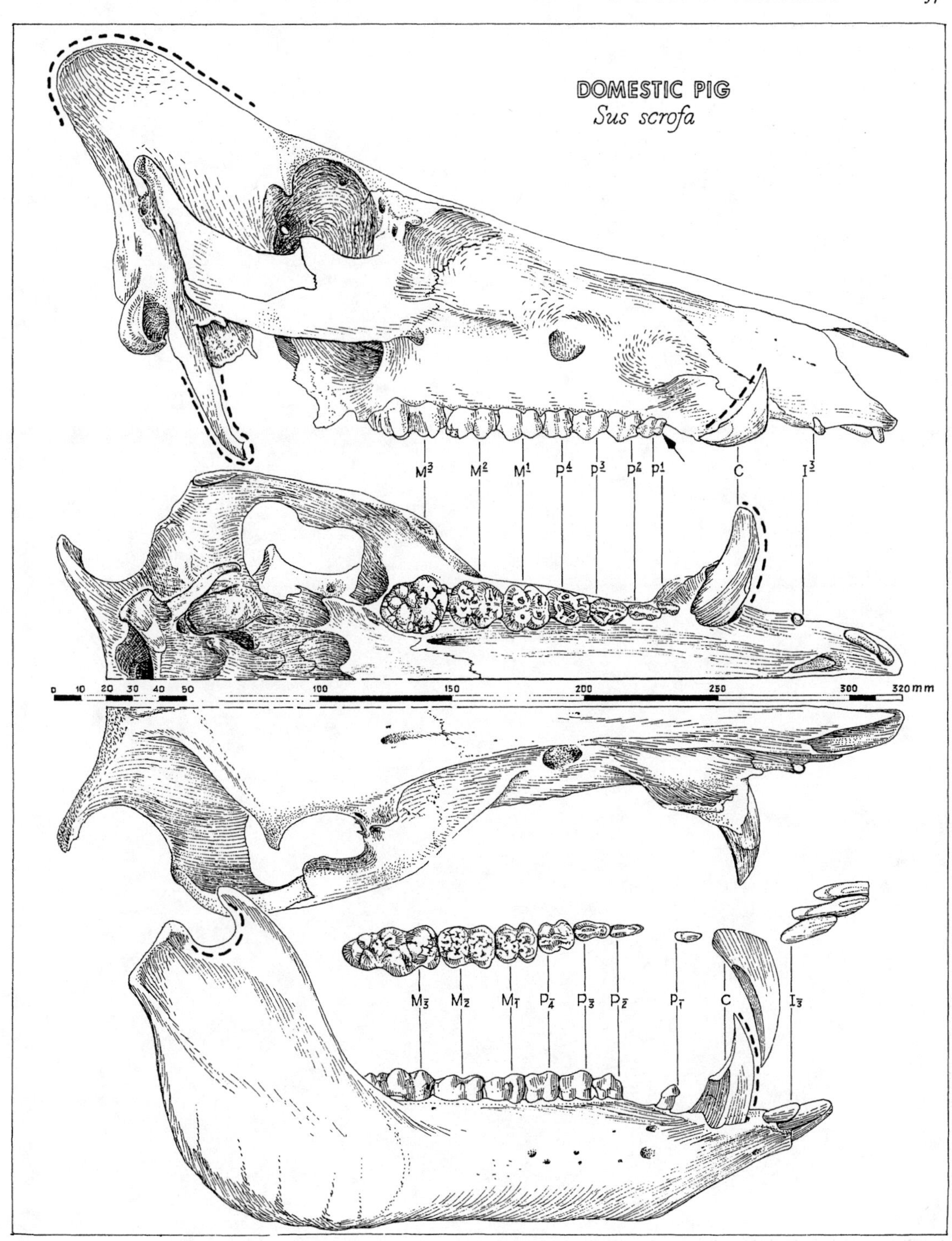
DOMESTIC PIG
Sus scrofa
M3 M2 M1 P4 P3 P2 P1 C I3
0 10 20 30 40 50 100 150 200 250 300 320 mm
M3 M2 M1 P4 P3 P2 P1 C I3

FIGURE 34

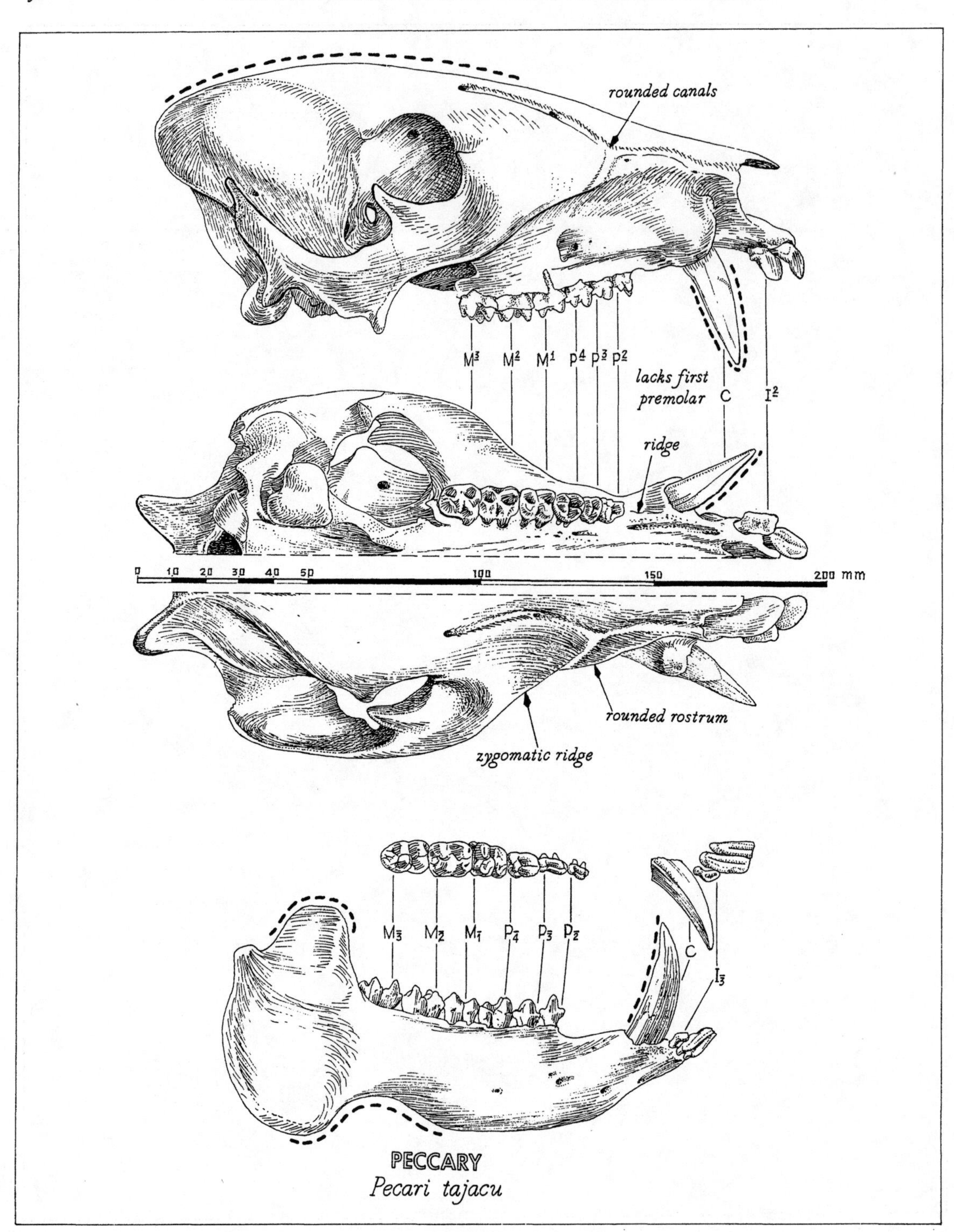
rounded canals
M3 M2 M1 P4 P3 P2
lacks first premolar
C
I2
ridge
0 10 20 30 40 50 100 150 200 mm
rounded rostrum
zygomatic ridge
M3 M2 M1 P4 P3 P2
C
I3
PECCARY
Pecari tajacu

FIGURE 35

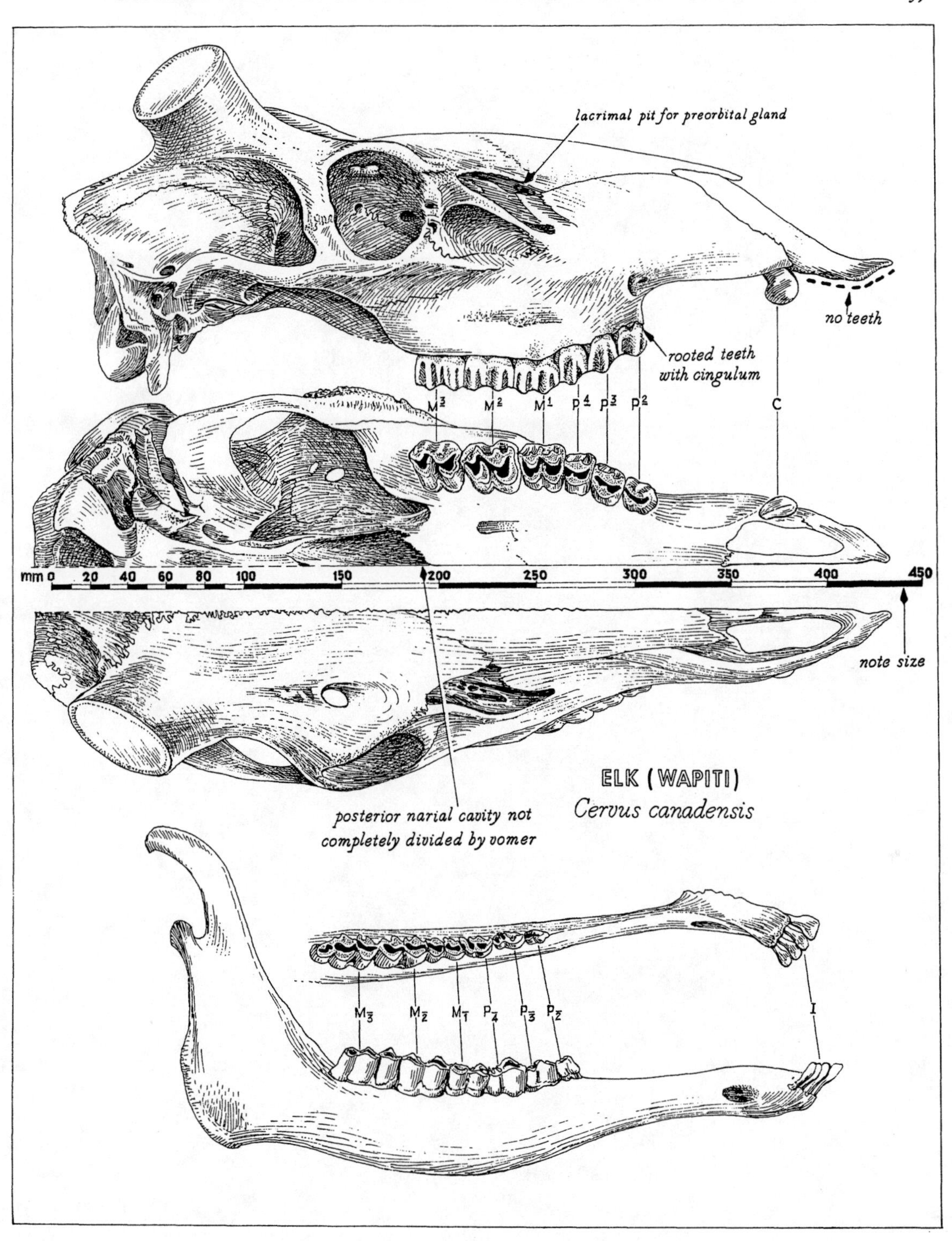
lacrimal pit for preorbital gland
no teeth
rooted teeth
with cingulum
M3 M2 M1 P4 P3 P2
C
mm 0 20 40 60 80 100 150 200 250 300 350 400 450
note size
ELK (WAPITI)
Cervus canadensis
posterior narial cavity not
completely divided by vomer
M3 M2 M1 P4 P3 P2
I

FIGURE 36

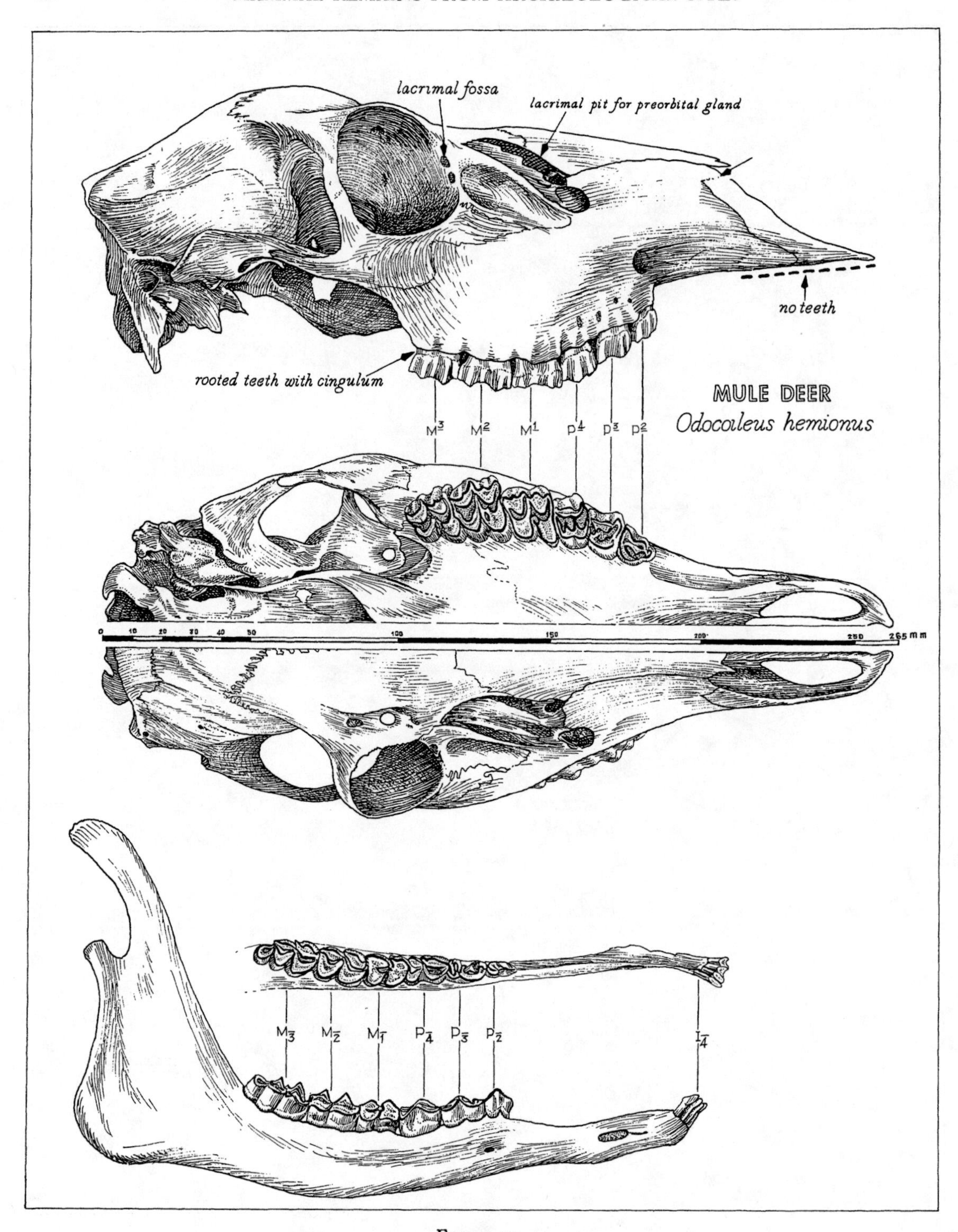
lacrimal fossa
lacrimal pit for preorbital gland
no teeth
rooted teeth with cingulum
MULE DEER
Odocoileus hemionus
M3 M2 M1 P4 P3 P2
0 10 20 30 40 50 100 150 200 250 265 mm
M3 M2 M1 P4 P3 P2 I4

FIGURE 37

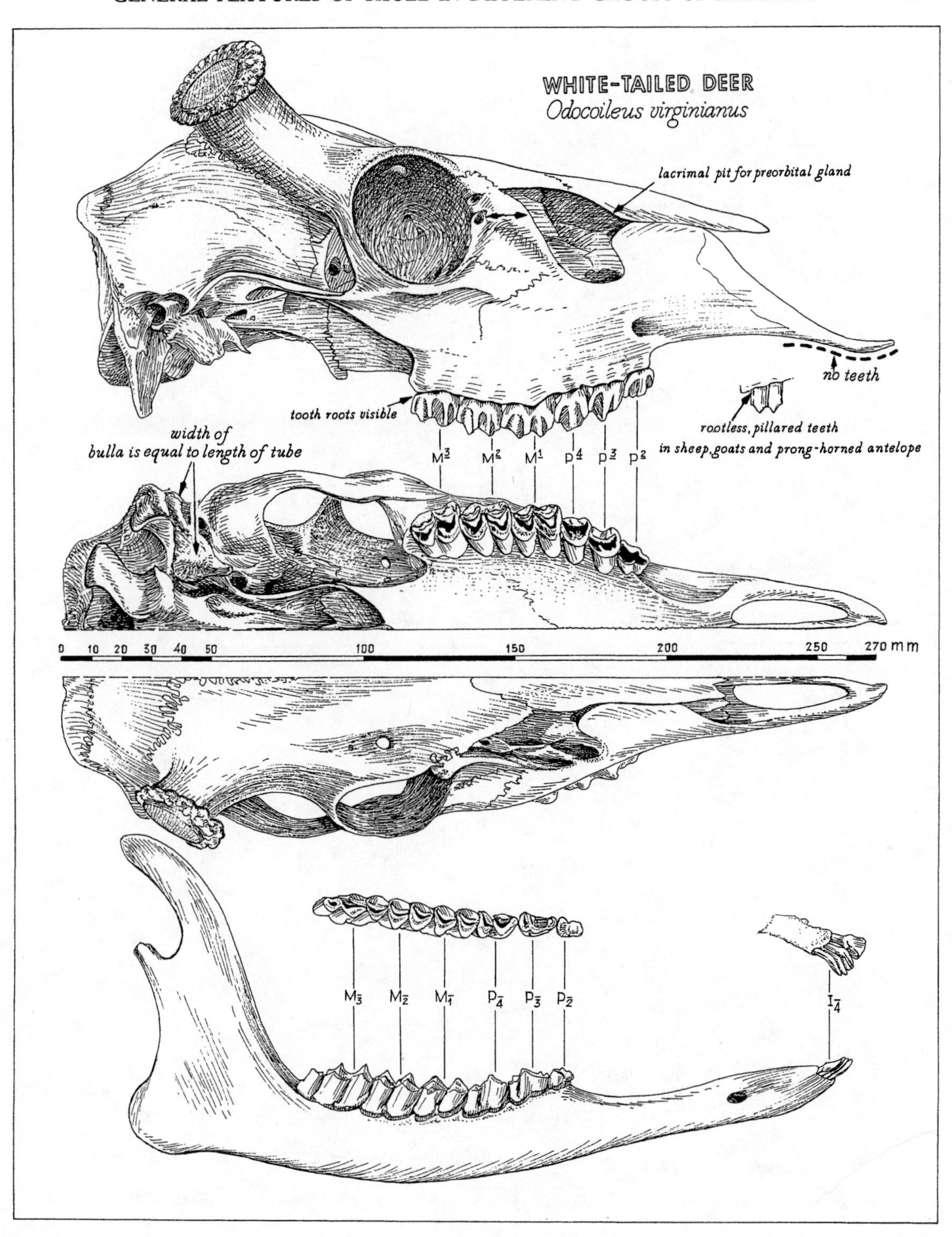
WHITE-TAILED DEER
Odocoileus virginianus
lacrimal pit for preorbital gland
no teeth
tooth roots visible
rootless, pillared teeth
in sheep, goats and prong-horned antelope
width of
bulla is equal to length of tube
M^3 M^2 M^1 P^4 P^3 P^2
0 10 20 30 40 50 100 150 200 250 270 mm
M_3 M_2 M_1 P_4 P_3 P_2
I_4

FIGURE 38

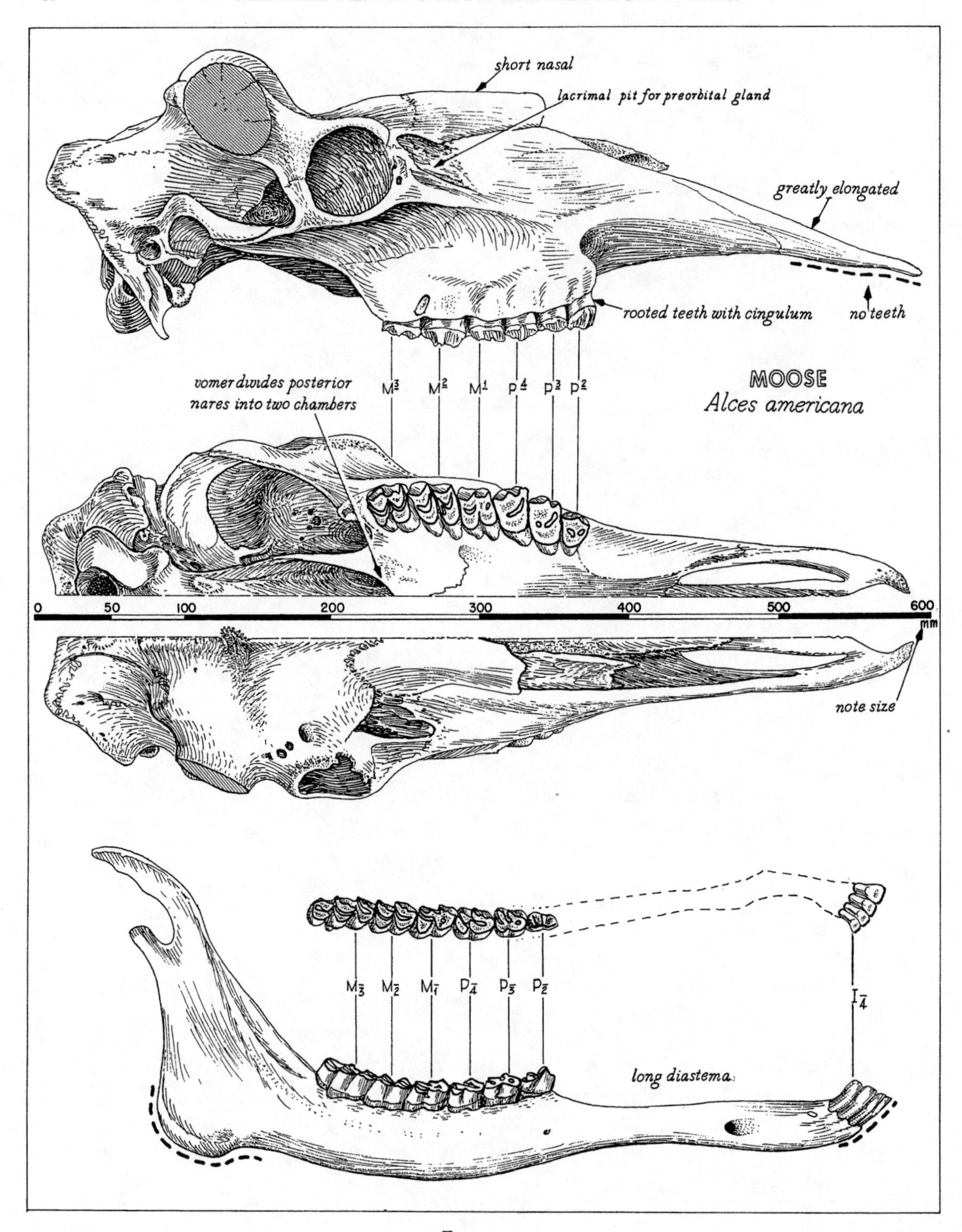
short nasal
lacrimal pit for preorbital gland
greatly elongated
rooted teeth with cingulum
no teeth
MOOSE
Alces americana
vomer divides posterior nares into two chambers
M3 M2 M1 P4 P3 P2
0 50 100 200 300 400 500 600
mm
note size
M3 M2 M1 P4 P3 P2
I4
long diastema

FIGURE 39

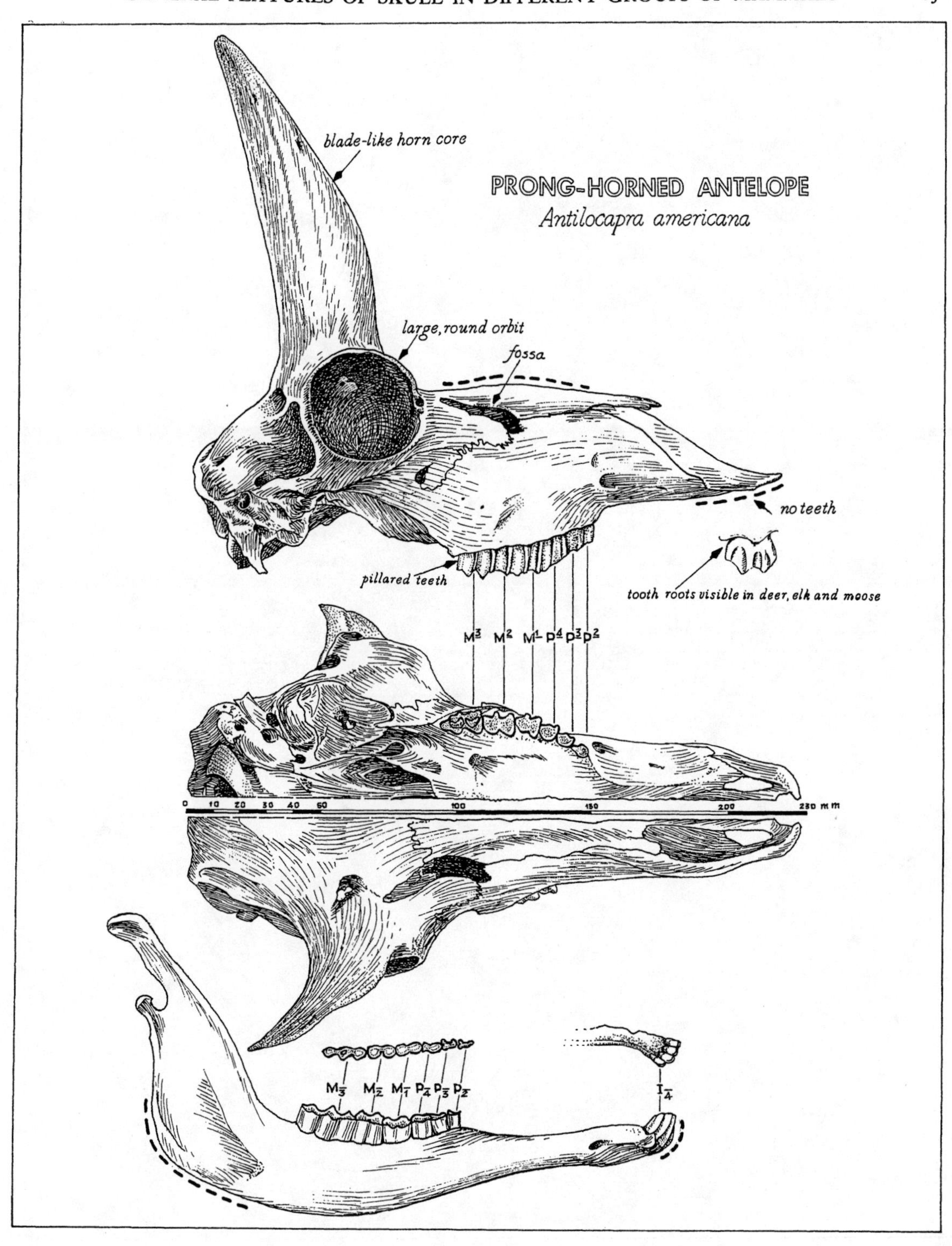
blade-like horn core
PRONG-HORNED ANTELOPE
Antilocapra americana
large, round orbit
fossa
no teeth
pillared teeth
tooth roots visible in deer, elk and moose
M3 M2 M1 P4 P3 P2
0 10 20 30 40 50 100 150 200 230 mm
M3 M2 M1 P4 P3 P2
I4

FIGURE 40

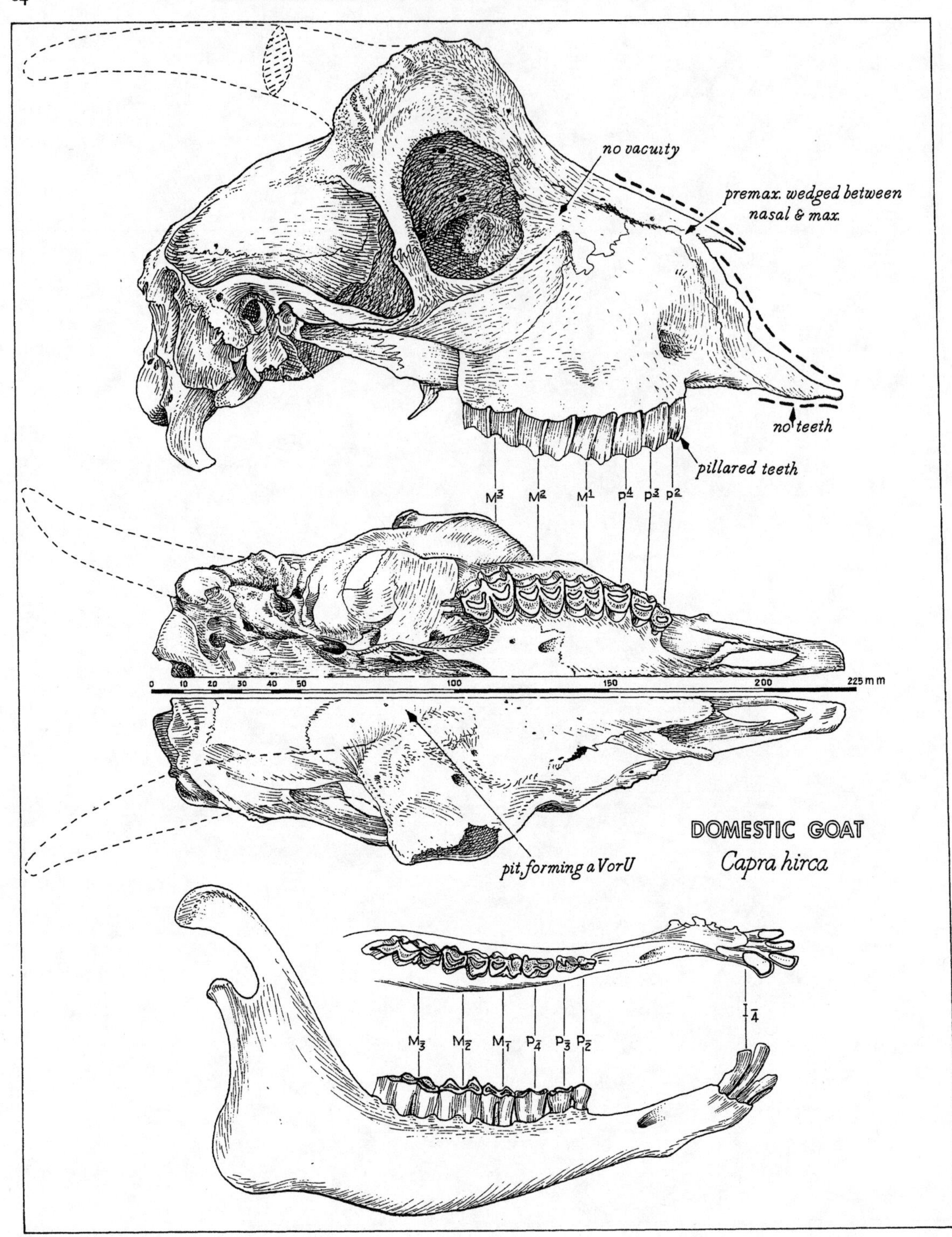
no vacuity
premax. wedged between
nasal & max.
no teeth
pillared teeth
M3 M2 M1 P4 P3 P2
0 10 20 30 40 50 100 150 200 225 mm
pit, forming a V or U
DOMESTIC GOAT
Capra hirca
M3 M2 M1 P4 P3 P2
I4

FIGURE 41

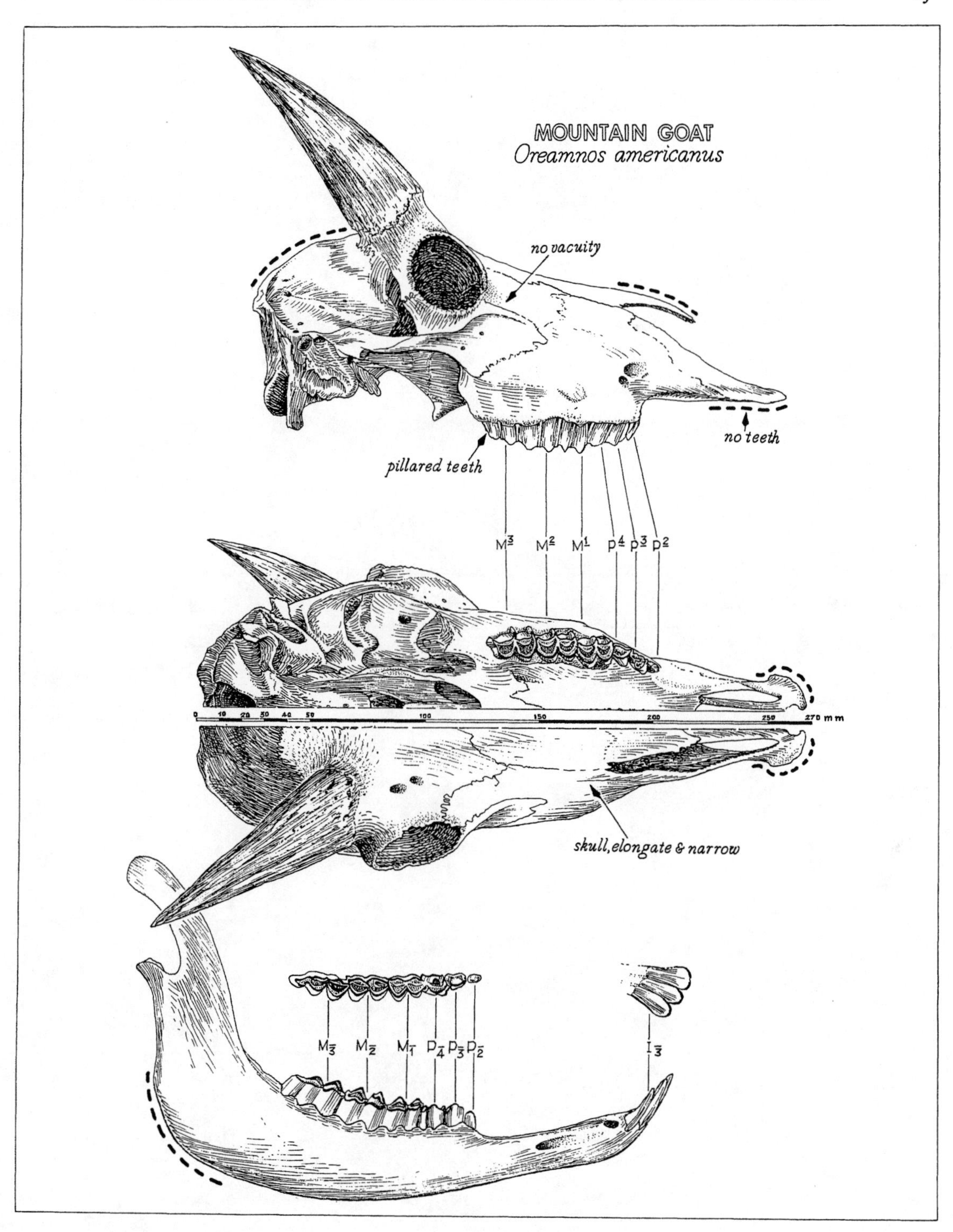
MOUNTAIN GOAT
Oreamnos americanus
no vacuity
no teeth
pillared teeth
M3 M2 M1 P4 P3 P2
0 10 20 30 40 50 100 150 200 250 270 mm
skull, elongate & narrow
M3 M2 M1 P4 P3 P2
I3

FIGURE 42

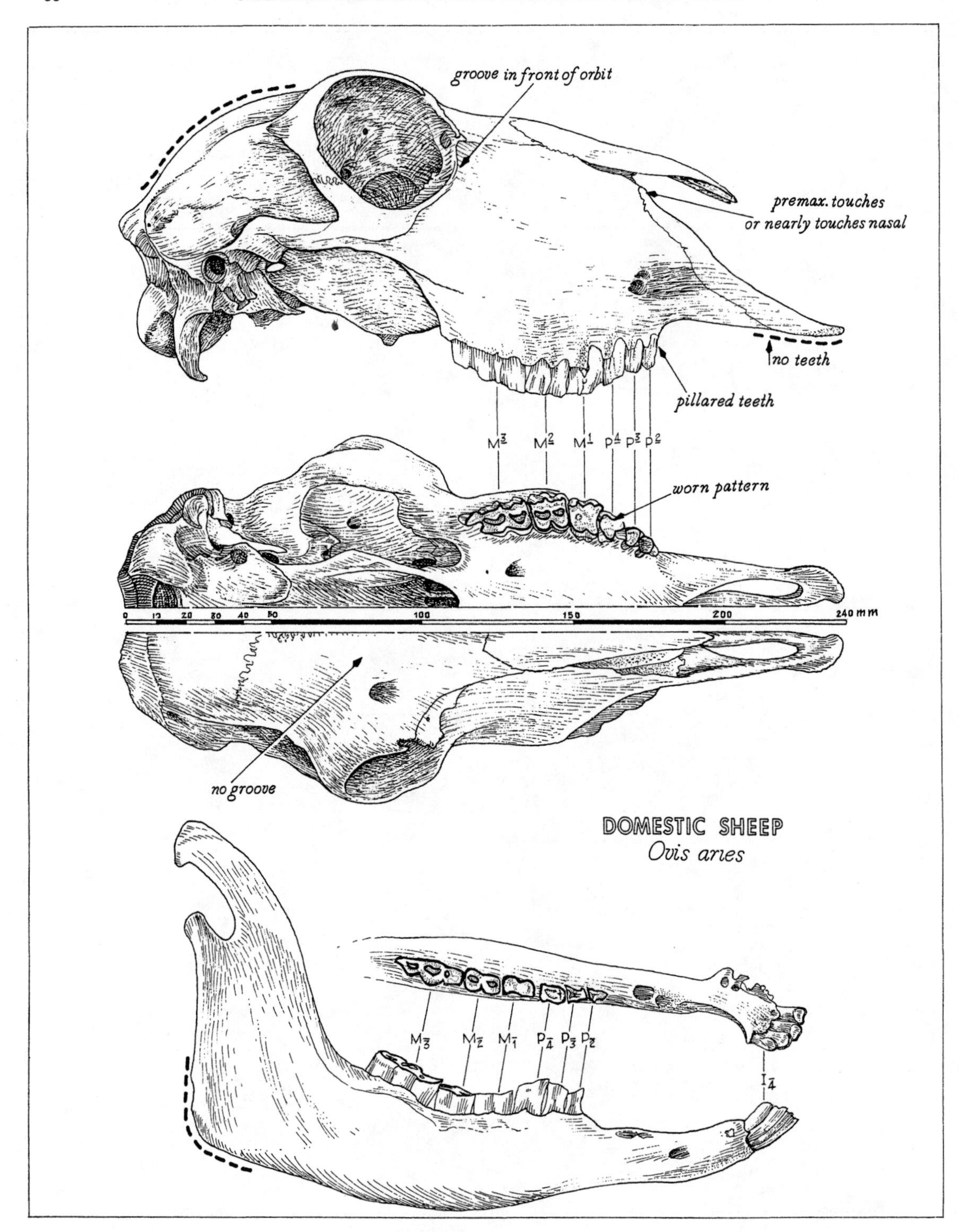
groove in front of orbit
premax. touches
or nearly touches nasal
no teeth
pillared teeth
M3 M2 M1 P4 P3 P2
worn pattern
0 10 20 30 40 50 100 150 200 240 mm
no groove
DOMESTIC SHEEP
Ovis aries
M3 M2 M1 P4 P3 P2
I4

FIGURE 43

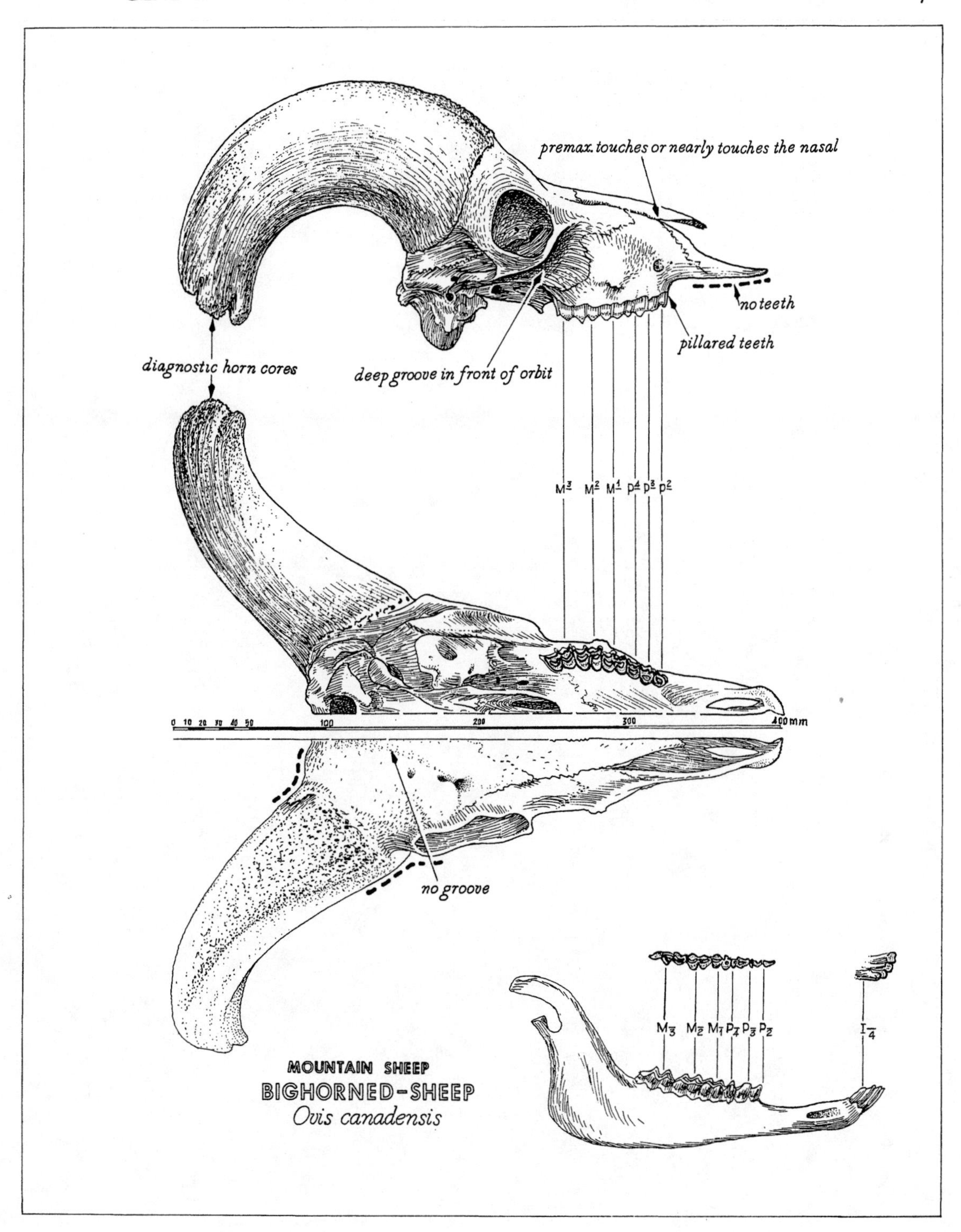
premax. touches or nearly touches the nasal
no teeth
pillared teeth
diagnostic horn cores
deep groove in front of orbit
M3 M2 M1 P4 P3 P2
0 10 20 30 40 50 100 200 300 400 mm
no groove
M3 M2 M1 P4 P3 P2
I4
MOUNTAIN SHEEP
BIGHORNED-SHEEP
Ovis canadensis

FIGURE 44

THE POSTCRANIAL SKELETON OF THE MAMMAL

THERE are many borderline cases, relating to the identification of postcranial elements, where it cannot be stated with certainty whether the bone in question is from a wild form or from an introduced species. This is particularly true in the goat-sheep group. Unless single bones can be separated with certainty, the published keys of previous workers are not carried over in this contribution.

Generic differences are usually well established in the skeleton. Some few species can be identified by isolated single bones. This is true for the armadillo and opossum, but these forms represent genera having but one or two species.

The mammal skeleton possesses features that are peculiar to this group alone. The ribs of the neck are fused to the vertebrae to form integral parts of these bones, whereas in the posterior part of the vertebral column the lumbar vertebrae are free of ribs. There is a marked distinction in the form of the cervical, thoracic, lumbar, sacral and caudal vertebrae. A strong spine is usually present on the blade of the scapula. The bones of the pelvis, ilium, ischium and pubis are fused into a single bony structure. An epiphysis is present on certain bones which, after fusion, indicates that bone growth is over; in contrast, reptilian bones reach no definite adult size, due to continuous growth.

Some elements of the postcranial skeleton are more diagnostic than others in determining which species are present. This subject has been discussed in a previous publication (Olsen, 1961a).

Many families are conservative and uniform as regards postcranial skeletons. In many instances one genus cannot be separated from another by any one bone if only morphological characters (not size) are considered. The skeletons illustrated in the following section are those of adult animals and it cannot be too strongly emphasized that close attention should be paid to the size of the element being compared. In some canids, for example, this may be the only means of separating similar-appearing forms.

The bacula are diagnostic taxonomic elements and are discussed in some detail in a separate publication (Burt, 1960). The same may be said for use of the calcaneum in taxonomic studies (Stains, 1959).

Many species of animals are represented in museum collections by one or two skeletons, although there may be several hundred skulls in the collection (this is particularly true of rodents in general). In very few instances was it noted that the sex of the animal was entered in the museum's osteological records or attached to the labeled skeleton, thereby making it impossible to record any size variation that might be attributed to the sex of the individual.

FIGURE 45

Generalized dog skeleton showing position of individual bones in articulation.

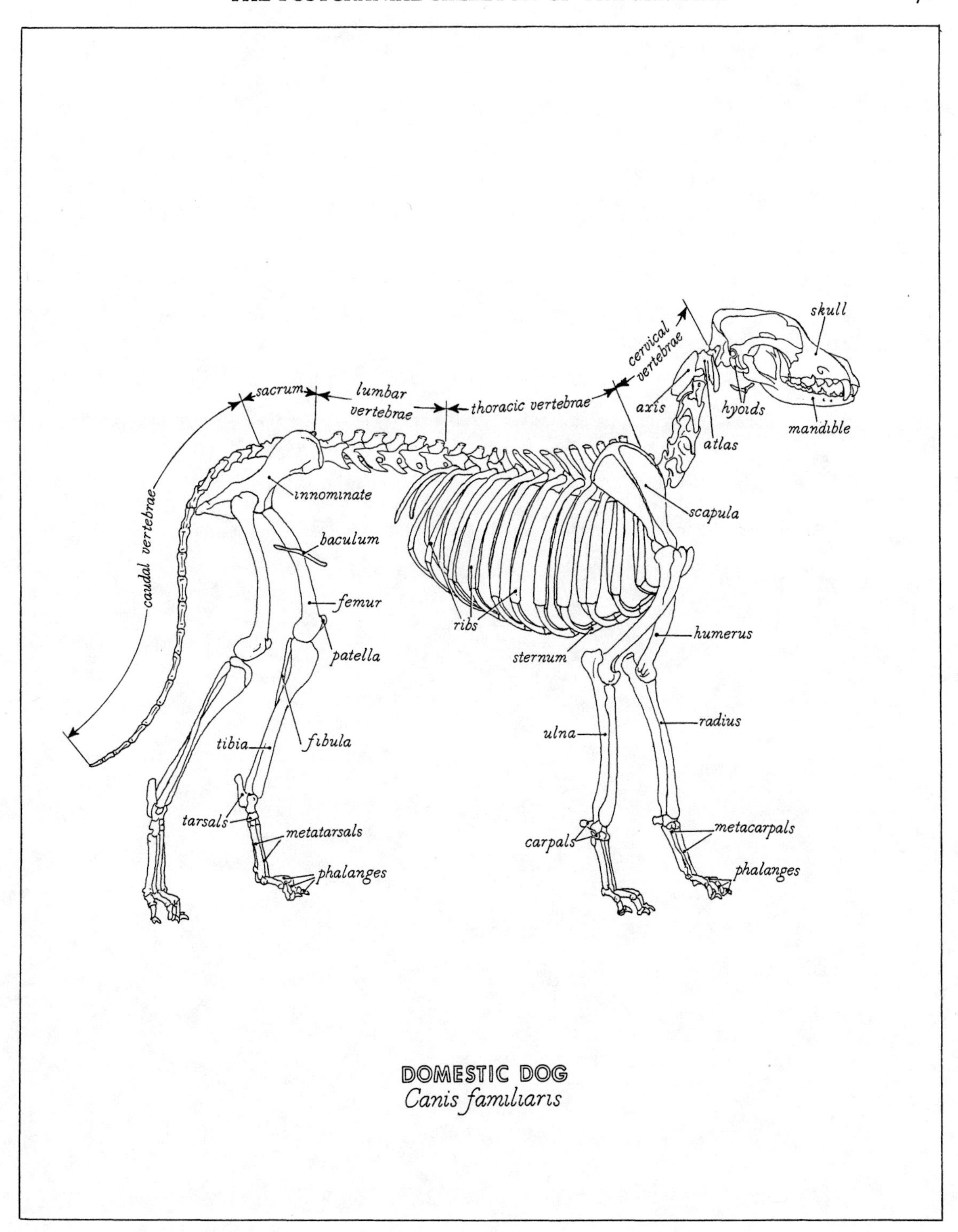

DOMESTIC DOG
Canis familiaris

FIGURE 46

Articulated fore limb of generalized dog showing relationship of the various elements which make up this structure.

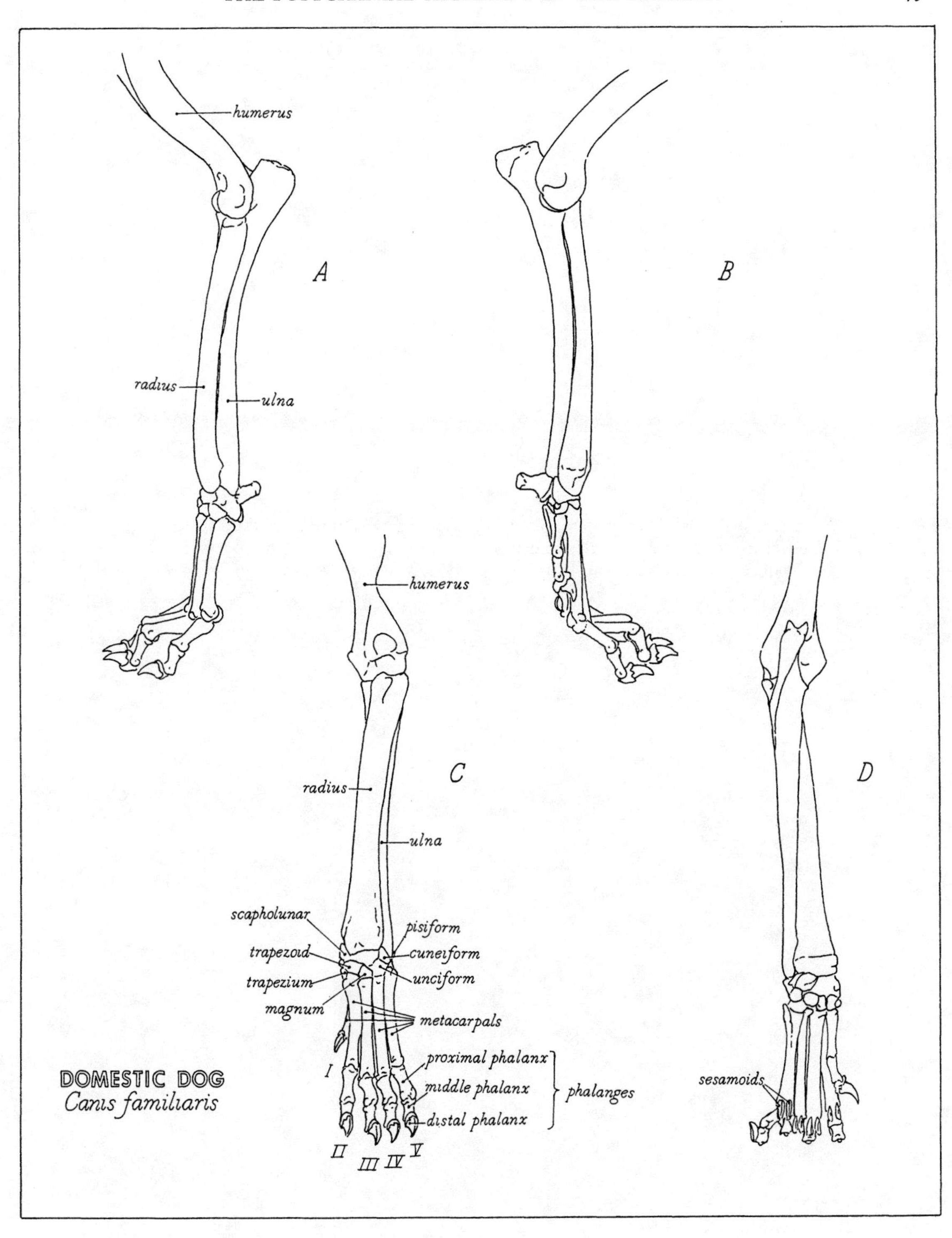
humerus
A
radius
ulna
B
humerus
C
radius
ulna
scapholunar
trapezoid
trapezium
magnum
pisiform
cuneiform
unciform
metacarpals
proximal phalanx
middle phalanx
distal phalanx
phalanges
I
II
III
IV
V
D
sesamoids
DOMESTIC DOG
Canis familiaris

FIGURE 47

Articulated hind limb of generalized dog showing relationship of the various elements which make up this structure.

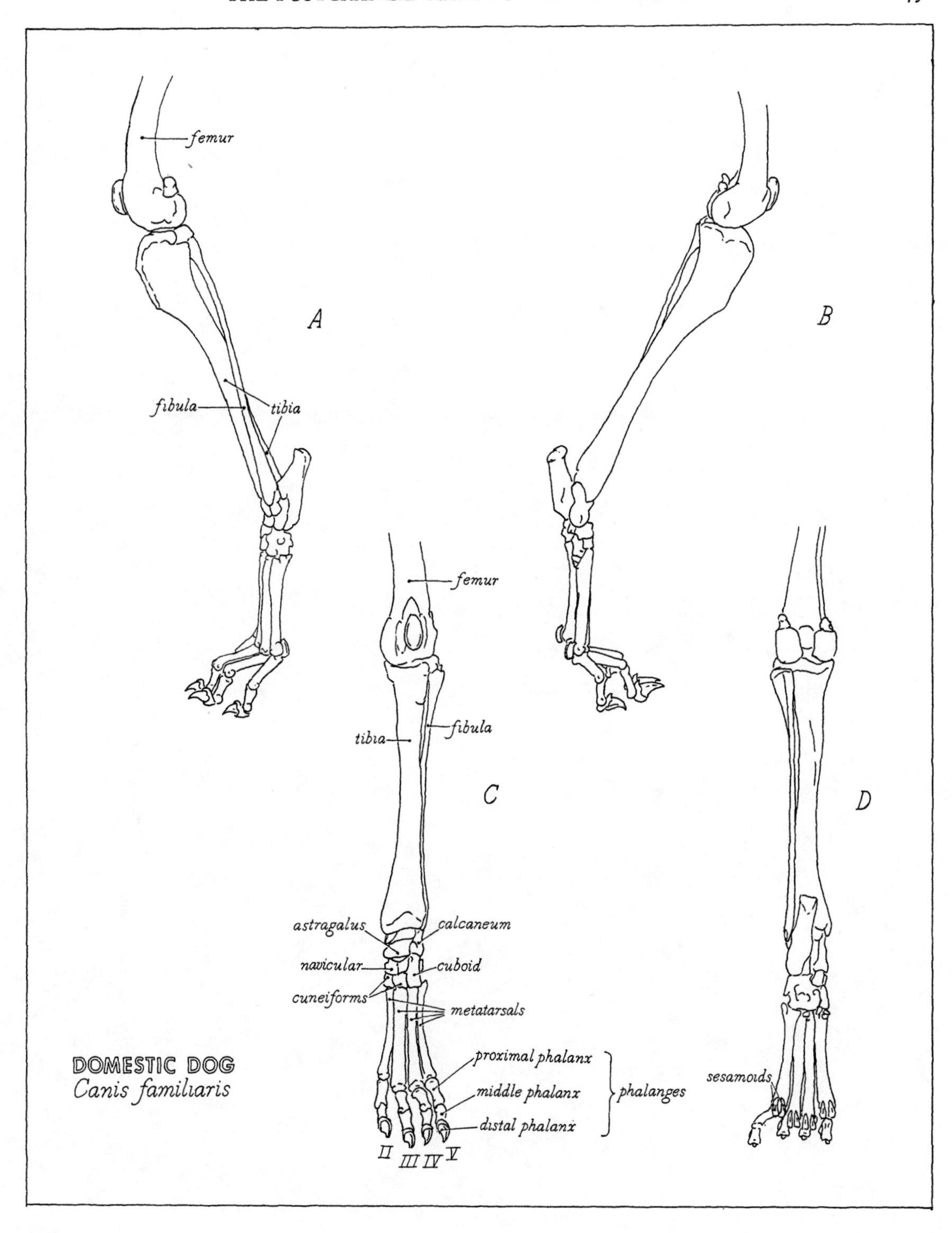
femur
A
fibula
tibia
B
femur
tibia
fibula
C
D
astragalus
calcaneum
navicular
cuboid
cuneiforms
metatarsals
proximal phalanx
middle phalanx
distal phalanx
phalanges
sesamoids
II
III IV V
DOMESTIC DOG
Canis familiaris

GENERAL FEATURES OF THE SKELETON IN DIFFERENT GROUPS OF MAMMALS

(Specific differentiating characters are indicated or noted on the plates depicting the individual bones.)

Marsupialia

Didelphiidae: Only one marsupial is considered, the opossum *Didelphis marsupialis.* Its key characters are indicated on the plates illustrating this animal's skeleton.

Edentata

Dasypodidae: Only one edentate is considered, the armadillo, *Dasypus novemcinctus.* Its key characters are indicated on the plates illustrating this animal's skeleton.

Lagomorpha

The leporid femur is long and slender with the third trochanter immediately below the greater trochanter. The tibia and fibula are fused at the distal end. The hind legs are longer than the front legs in rabbits. The ulna is reduced along the middle of the shaft in the jackrabbit and, excepting the lower extremity, is placed almost entirely behind the radius. In the desert cottontail the radius and ulna are of about equal size. The scapula is long and narrow with a long and narrow acromion and a prominent metacromion process. There is a deep notch between the acromion and the spine. In the ochotonids the fibula is fused as in the rabbits. The hind limbs are scarcely longer than the forelimbs (as compared with rabbits.) There is no pubic symphysis.

Rodentia

The sacrum in rodents generally consists of a single wide vertebra connecting the ilia; the others are narrow and ankylosed to it. The scapula is often long and narrow with a variable outline. The acromion is long and narrow, often with a large metacromial process. Some species have a deep notch between the acromion process and the spine. The coracoid is a small blunt process. All bones are slight and slender. The ilium is rod-like and scarcely expanded dorsally. The pubis and ischium are long, diverging caudally. The symphysis is long and generally ossified. The humerus is generally slender and straight with slightly developed muscular processes. There is at times a supratrochlear perforation and generally no entepicondylar foramen. The radius and ulna are separate. The femur generally has a distinct neck and often a small head. The distal ends of the tibia and fibula are fused in rats and mice. The sciurid femur is long and slender with a third trochanter immediately below the greater trochanter. The tibia and fibula are never fully fused. In the Castoridae the sacral vertebrae have increasingly wide transverse processes which nearly meet the ischia. The scapula has a large, strong but undivided acromion. The femur is short and stout with a well-developed third trochanter at midshaft. The tibia and fibula are separate. In the cricetids the tibia and fibula are united at the distal ends. In the zapodids the tibia and fibula are fused at the distal ends. The femur has no sign of a third trochanter in the erethizontids; the great trochanter is very large and stands even with or above the level of the head of the femur. The tibia and fibula are separate, or at times fused at both ends in old individuals.

Carnivora

The scapula of the carnivores has a well-developed spine dividing the blade into nearly equal infraspinous and supraspinous fossae. The acromion is prominent, sometimes with a well-marked metacromial process. The coracoid is small. The pelvis is long and narrow and more or less parallel-sided with its ventral side not everted. The symphysis is long and formed in part by the ischia as well as by the pubes. The humerus has a strongly curved shaft with a deltoid ridge. The ulna has a long transversally compressed olecranon process. The femur is long and slender and straight. The great trochanter is seldom higher than the head except in the felids. There is no third trochanter. The tibia and fibula are distinct and generally separate.

The canids have the following additional characteristics: The scapula is subrectangular, except in the fox where it is rounded, with a marked corocoid notch. The spine is located diagonally. The acromion is more pronounced than in the felids. The metacromion is only slightly present. The humerus has a large supratrochlear perforation. There is no entepicondylar foramen. The lower half of the fibula lies close alongside the tibia.

The ursids have no entepicondylar foramen in the humerus. The scapula is "D" shaped and has a prominent postscapular fossa (as does the raccoon). There is a strong acromion with no indication of a metacromion process. There is a prominent ridge, laterally below the greater trochanter.

It is possible for some elements of the bear skeleton to be misidentified as belonging to a human. The fact that this is of some concern is apparent in the FBI bulletin (Stewart, 1959), which discusses the similarities between the bones of the human hand and those of the ursid paw. The long bones of the limbs also have occasionally been attributed to those of man.

The mustelids have a scapular outline that is variable but it tends to be generally short and broad with a wide glenoid cavity, compared with the total area of the scapular blade. Most forms have an entepicondylar foramen in the humerus (excepting skunks).

The felids have a "D" shaped scapula. The caudal border is rectilinear; the cranial border is almost semicircular. The spine crosses the blade diagonally. There is a strong acromion with a short, stout metacromion process. The crest of the spine is turned caudally. The humerus has an entepicondylar foramen. The great trochanter is higher than the head of the femur.

Perissodactyla

Equidae: Only the horse, *Equus caballus*, is considered under the perissodactyls. The scapula is long, narrow, and somewhat triangular, and the spine divides the blade unequally. There is a rudimentary acromion. The coracoid is a prominent knob. The spine ends in the midline of the neck. There is a strong third trochanter present on the femur. The femur is very strongly constructed; the greater trochanter is enormous and extends above the head. There is no constricting neck to the femur. The lesser trochanter is a low, roughened ridge located on the medial face of the shaft. The ilium is broad, but contracts to a long, slender neck above the acetabulum. The ischium is short. The pubic symphysis is short. The obturator foramen is circular in shape. The acetabular notch is wide open, ventrally. The humerus has well-defined tuberosities and a marked deltoid ridge. The ulna is fused to the radius at midshaft. The radius is strong, short and heavily constructed, being rather "D" shaped in cross section. The fibula is much reduced, and consists of a mere splint with a roughened head. The shaft tapers to a sharp point near the middle of the tibia and forms a natural awl. (A word of caution is given here, in the event that the shape of an isolated horse fibula be attributed to the handiwork of man.)

Artiodactyla

The artiodactyl skeleton has a considerable number of distinguishing features. There are two terminal phalanges on each foot giving them a "cloven foot". The two principal metapodials fuse together into a single cannon-bone. There is no third trochanter on the femur. The ulna is reduced or fused to the radius. The astragalus forms a characteristic double pulley. The entire pelvis is elongated. The ilium is relatively short and expanded and everted dorsally. The ischium is long, wide and everted caudally. There is a well-defined tuberosity in the center of the caudal extremity. The symphysis is long and the obturator foramen forms a long oval. The humerus has no entepicondylar foramen. There is no constricting neck to the femur. There is a reduction of the fibula until nothing remains but its lower extremity and a vestige of the proximal end.

The tayassuid skeleton has a scapula that forms an isosceles triangle, having regular borders. The spine divides the blade into nearly equal areas. The spine ends without an acromion near the midline of the neck. The crest of the spine is turned along most of its length to overhang the scapular fossa. There is a well-marked midspinous process in its dorsal one-third. The coronoid process is present as a swelling, cranial to the glenoid cavity. The neck is well marked but with no coracoid notch. The acetabular notch is wide open ventrally as in the perissodactyls. The humerus is short and heavy with a small head and tuberosities that are prominent and overhang the bicipital groove. The deltoid ridge is prominent. The supinator ridge runs out to the lateral epicondyle. The olecranon fossa is short and oval-shaped. The trochlea is narrow. The radius and ulna are separate. The limb bones are proportionately longer and more slender in the wild pigs. The femur has a small round head, terminating in a comparatively long neck. The greater trochanter is large and barely exceeds the head in height. The lesser trochanter is present as a stout knob. No third trochanter is present.

The cervids have a scapula, almost like an isosceles triangle in shape, having a spine that arises at the margin between the cranial one-fourth and the caudal three-fourths of the dorsal border, and runs in an almost straight line to the neck of the scapula where it terminates well short of the glenoid cavity. The acromion is present as a slight, thickened projection, separated by a shallow notch from the terminus of the spine at the neck of the blade. The coracoid is short and hooked with a slight rudimentary process. The acetabular notch of the pelvis is more or less closed over near the ventral margin of the acetabulum by two overhanging edges of the articular surfaces. The humerus is short and stout. The radius and ulna are long by comparison. The shaft of the ulna may be a thin splint-like bone. The radius is a curved, strong bone with a "D" shaped cross section. The flattened posterior face of the radius has a roughened area adjacent to the proximal epiphysis. The radial shaft has a long, roughened scar on its medial portion, to which the shaft of the ulna is attached. The head of the femur tends to make more of a right angle with the shaft than other comparable-sized forms. The distal condyles are large and are placed to the rear of the axis of the shaft. The shaft of the femur is long and slender in this group, as compared with a shorter and more heavily built bone in the bovidae.

The general shape of the scapula in all ruminants is much alike with very little in the way of characteristics to distinguish one from the other. The spine tends to be more sinuous in the sheep and goats than in the deer. The subacromial notch is a bit more pronounced in the goats than in the sheep. The pelvis of the bovids tends to be heavier in build than that found in the cervids. However, a large comparative series would be needed to make adequate comparisons of similar-sized forms. The limb bones of the large cervids are distinguished from the bovids by their being more "long-limbed" and of lighter build. The bovids are shorter-legged and more heavily built. The bovid trochanters and condyles tend to be more heavily and strongly constructed than those of the large cervids, but here again it is a matter of comparison with an adequate skeletal collection before exact determinations can be made from these closely allied forms.

FIGURES 48 TO 99

Postcranial skeletal elements of the cited forms. Differentiating characters are indicated by notes and arrows or by heavy dashed lines. Elements of the various bones, processes and articulations that are common to all forms, are labeled with straight lines without arrows.

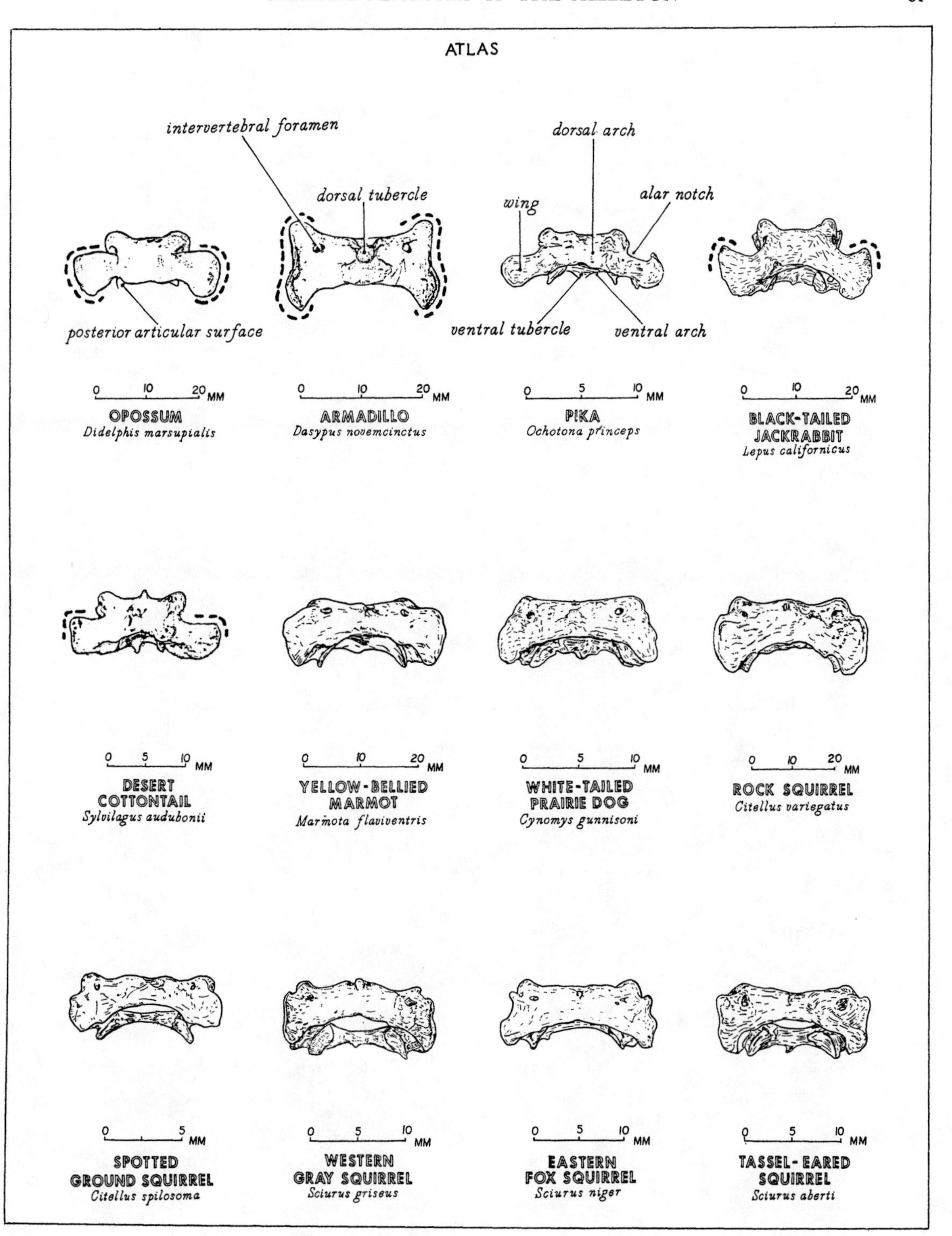
ATLAS
intervertebral foramen
dorsal tubercle
dorsal arch
wing
alar notch
posterior articular surface
ventral tubercle
ventral arch
0 10 20 MM
OPOSSUM
Didelphis marsupialis
0 10 20 MM
ARMADILLO
Dasypus novemcinctus
0 5 10 MM
PIKA
Ochotona princeps
0 10 20 MM
BLACK-TAILED JACKRABBIT
Lepus californicus
0 5 10 MM
DESERT COTTONTAIL
Sylvilagus audubonii
0 10 20 MM
YELLOW-BELLIED MARMOT
Marmota flaviventris
0 5 10 MM
WHITE-TAILED PRAIRIE DOG
Cynomys gunnisoni
0 10 20 MM
ROCK SQUIRREL
Citellus variegatus
0 5 MM
SPOTTED GROUND SQUIRREL
Citellus spilosoma
0 5 10 MM
WESTERN GRAY SQUIRREL
Sciurus griseus
0 5 10 MM
EASTERN FOX SQUIRREL
Sciurus niger
0 5 10 MM
TASSEL-EARED SQUIRREL
Sciurus aberti

FIGURE 48

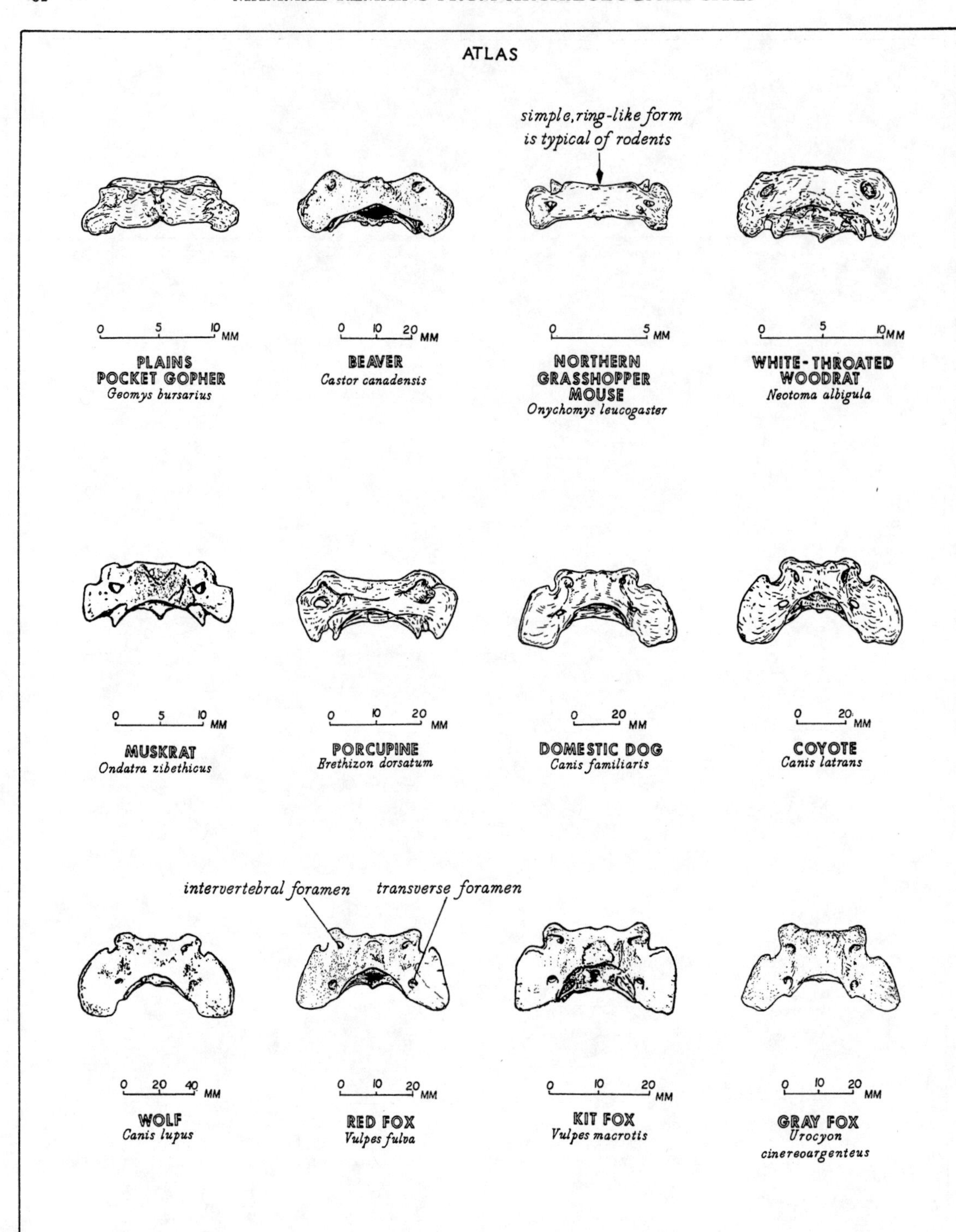
ATLAS
simple, ring-like form is typical of rodents
0 5 10 MM
PLAINS POCKET GOPHER
Geomys bursarius
0 10 20 MM
BEAVER
Castor canadensis
0 5 MM
NORTHERN GRASSHOPPER MOUSE
Onychomys leucogaster
0 5 10 MM
WHITE-THROATED WOODRAT
Neotoma albigula
0 5 10 MM
MUSKRAT
Ondatra zibethicus
0 10 20 MM
PORCUPINE
Erethizon dorsatum
0 20 MM
DOMESTIC DOG
Canis familiaris
0 20 MM
COYOTE
Canis latrans
intervertebral foramen
transverse foramen
0 20 40 MM
WOLF
Canis lupus
0 10 20 MM
RED FOX
Vulpes fulva
0 10 20 MM
KIT FOX
Vulpes macrotis
0 10 20 MM
GRAY FOX
Urocyon cinereoargenteus

FIGURE 49

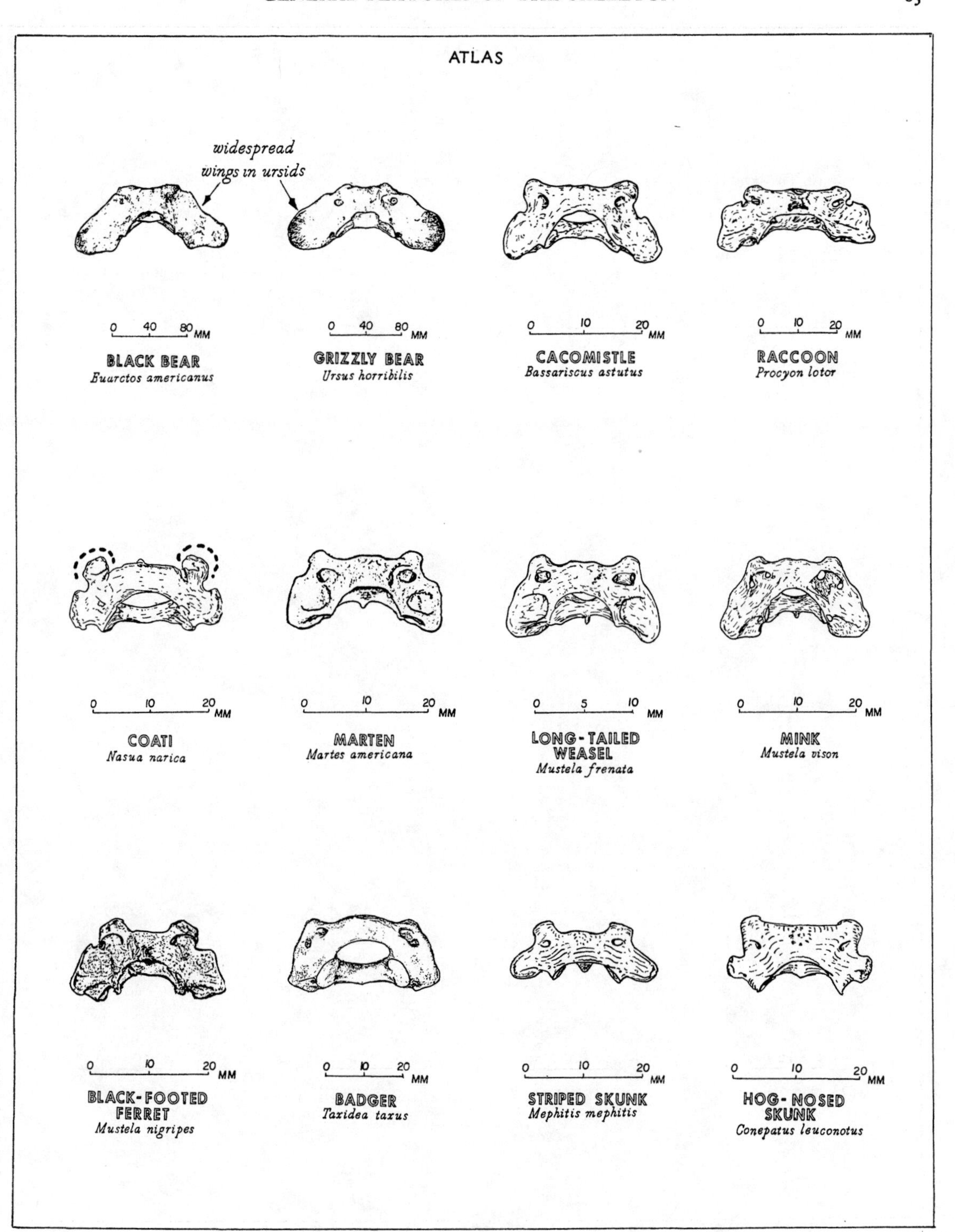
ATLAS
widespread
wings in ursids
0 40 80 MM
BLACK BEAR
Euarctos americanus
0 40 80 MM
GRIZZLY BEAR
Ursus horribilis
0 10 20 MM
CACOMISTLE
Bassariscus astutus
0 10 20 MM
RACCOON
Procyon lotor
0 10 20 MM
COATI
Nasua narica
0 10 20 MM
MARTEN
Martes americana
0 5 10 MM
LONG-TAILED
WEASEL
Mustela frenata
0 10 20 MM
MINK
Mustela vison
0 10 20 MM
BLACK-FOOTED
FERRET
Mustela nigripes
0 10 20 MM
BADGER
Taxidea taxus
0 10 20 MM
STRIPED SKUNK
Mephitis mephitis
0 10 20 MM
HOG-NOSED
SKUNK
Conepatus leuconotus

FIGURE 50

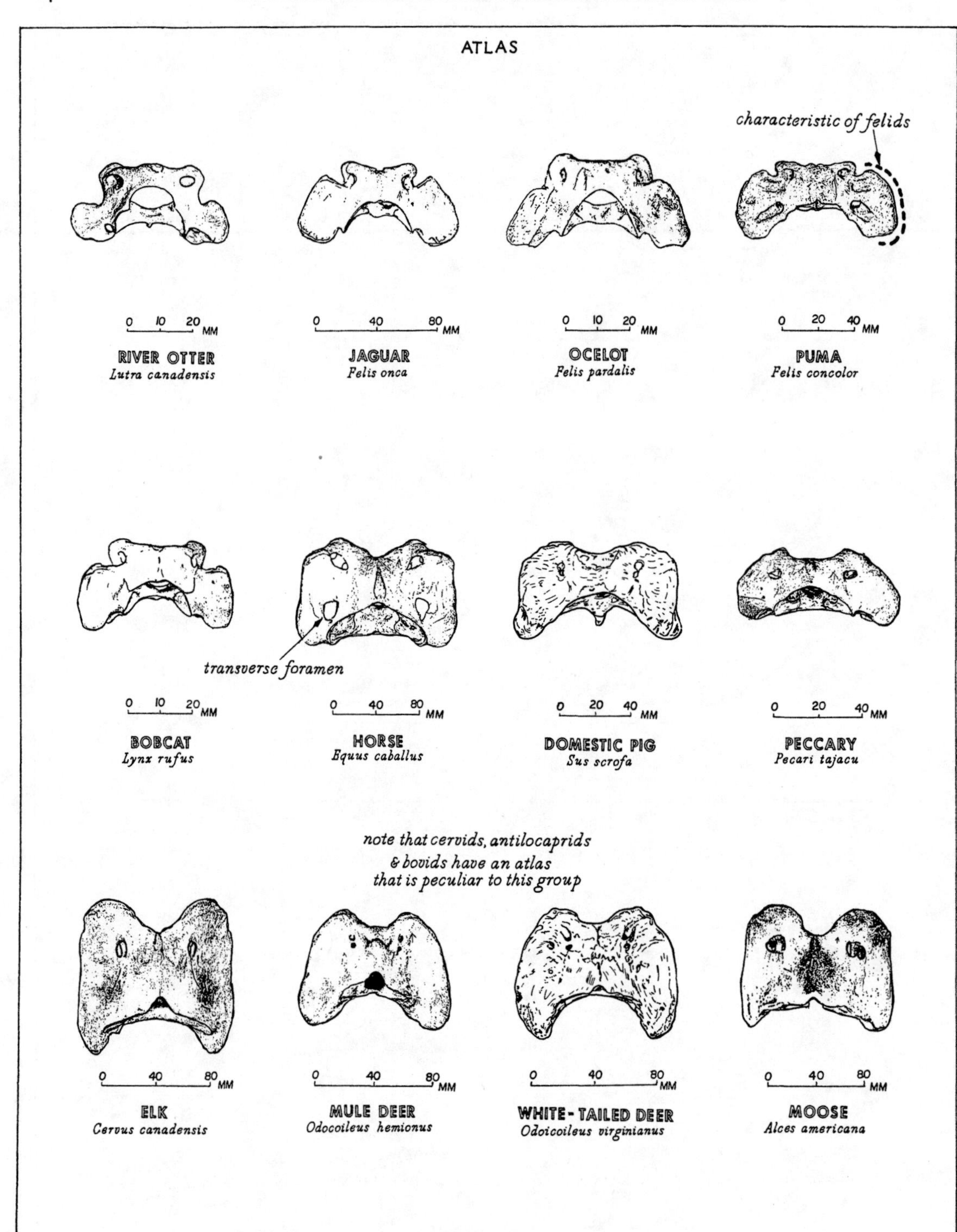
ATLAS
characteristic of felids
0 10 20 MM
RIVER OTTER
Lutra canadensis
0 40 80 MM
JAGUAR
Felis onca
0 10 20 MM
OCELOT
Felis pardalis
0 20 40 MM
PUMA
Felis concolor
transverse foramen
0 10 20 MM
BOBCAT
Lynx rufus
0 40 80 MM
HORSE
Equus caballus
0 20 40 MM
DOMESTIC PIG
Sus scrofa
0 20 40 MM
PECCARY
Pecari tajacu
note that cervids, antilocaprids
& bovids have an atlas
that is peculiar to this group
0 40 80 MM
ELK
Cervus canadensis
0 40 80 MM
MULE DEER
Odocoileus hemionus
0 40 80 MM
WHITE-TAILED DEER
Odoicoileus virginianus
0 40 80 MM
MOOSE
Alces americana

FIGURE 51

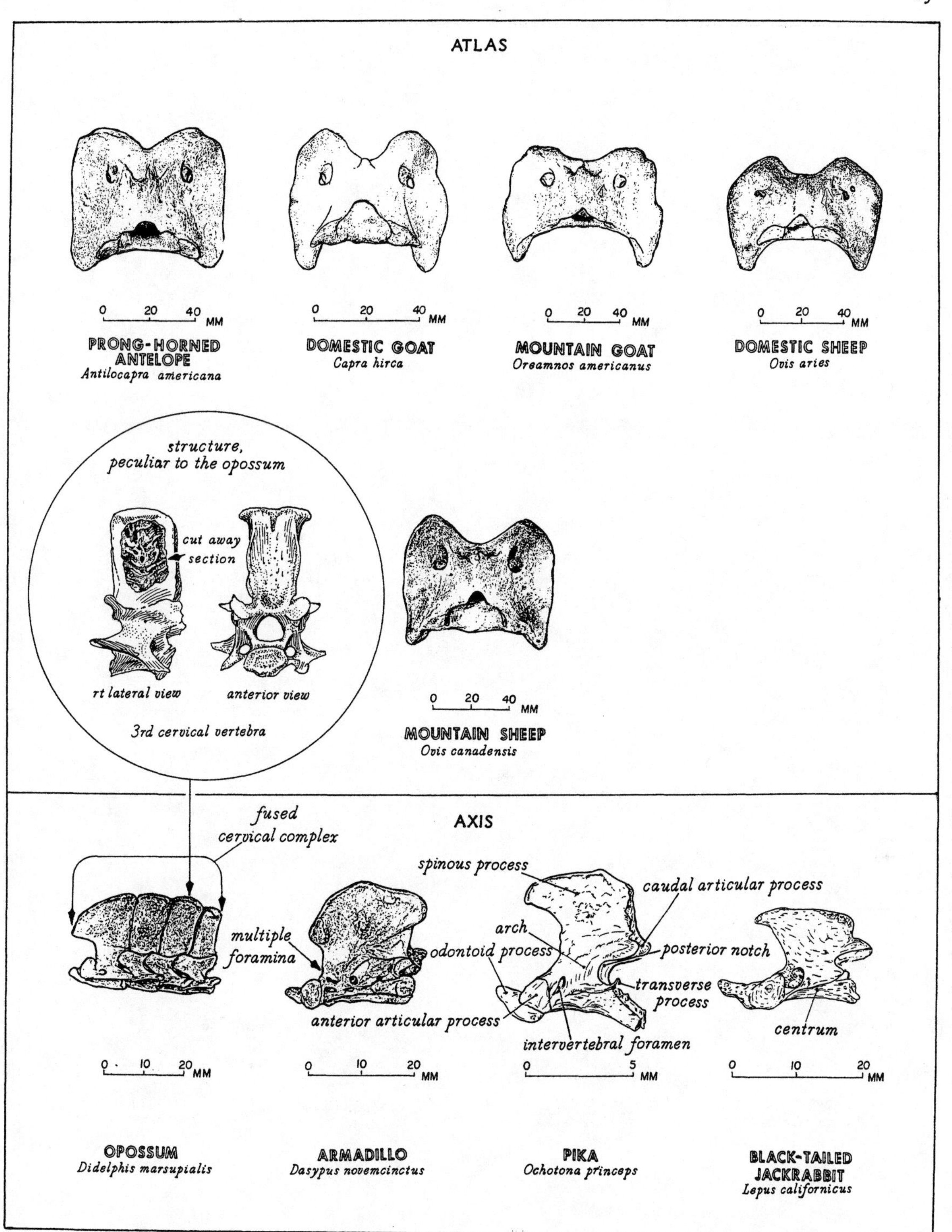
ATLAS
0 20 40 MM
PRONG-HORNED ANTELOPE
Antilocapra americana
0 20 40 MM
DOMESTIC GOAT
Capra hirca
0 20 40 MM
MOUNTAIN GOAT
Oreamnos americanus
0 20 40 MM
DOMESTIC SHEEP
Ovis aries
structure, peculiar to the opossum
cut away section
rt lateral view
anterior view
3rd cervical vertebra
0 20 40 MM
MOUNTAIN SHEEP
Ovis canadensis
AXIS
fused cervical complex
multiple foramina
spinous process
caudal articular process
arch
odontoid process
posterior notch
transverse process
anterior articular process
intervertebral foramen
centrum
0 10 20 MM
0 10 20 MM
0 5 MM
0 10 20 MM
OPOSSUM
Didelphis marsupialis
ARMADILLO
Dasypus novemcinctus
PIKA
Ochotona princeps
BLACK-TAILED JACKRABBIT
Lepus californicus

FIGURE 52

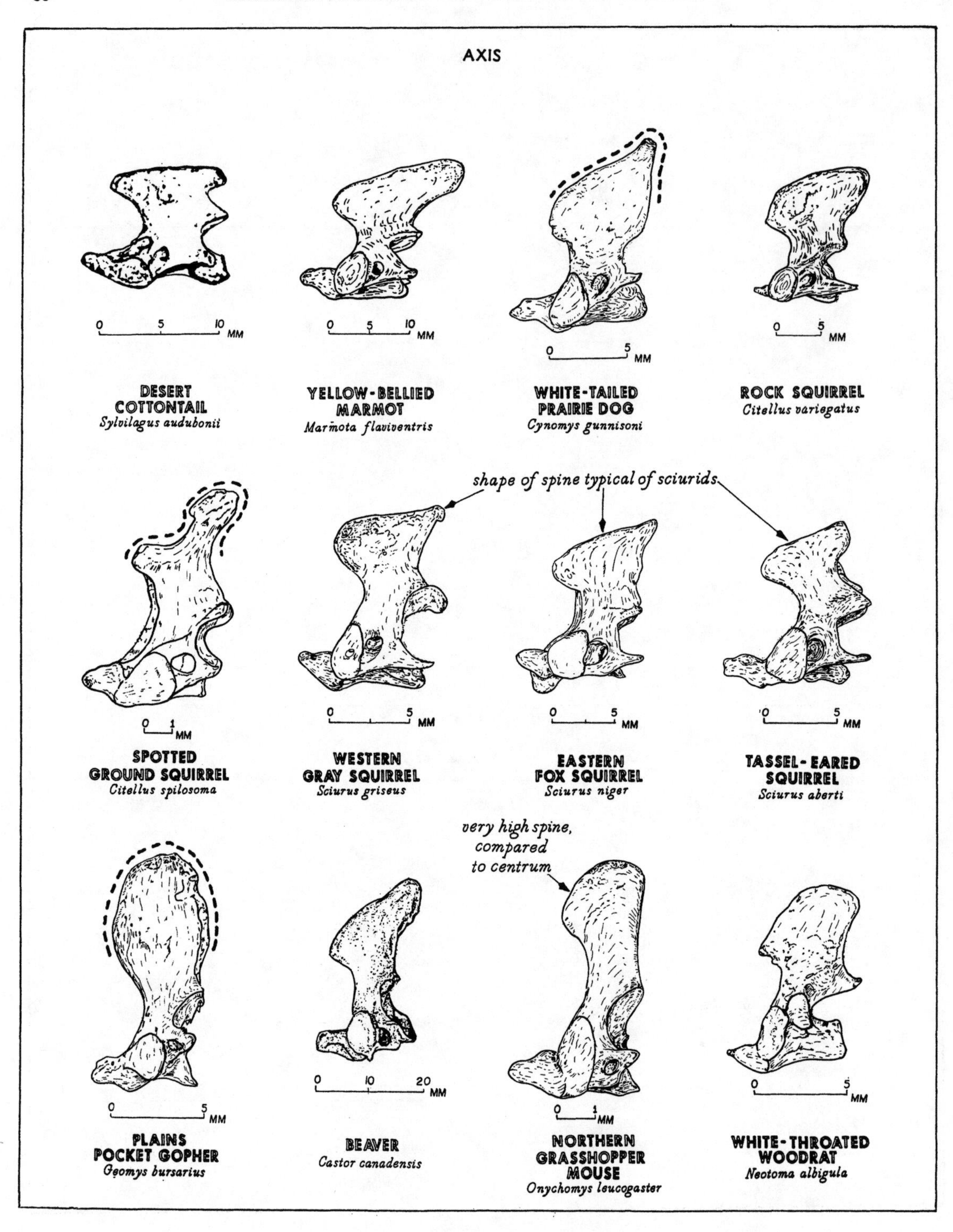

AXIS
0 5 10 MM
DESERT COTTONTAIL
Sylvilagus audubonii
0 5 10 MM
YELLOW-BELLIED MARMOT
Marmota flaviventris
0 5 MM
WHITE-TAILED PRAIRIE DOG
Cynomys gunnisoni
0 5 MM
ROCK SQUIRREL
Citellus variegatus
shape of spine typical of sciurids
0 1 MM
SPOTTED GROUND SQUIRREL
Citellus spilosoma
0 5 MM
WESTERN GRAY SQUIRREL
Sciurus griseus
0 5 MM
EASTERN FOX SQUIRREL
Sciurus niger
0 5 MM
TASSEL-EARED SQUIRREL
Sciurus aberti
very high spine, compared to centrum
0 5 MM
PLAINS POCKET GOPHER
Geomys bursarius
0 10 20 MM
BEAVER
Castor canadensis
0 1 MM
NORTHERN GRASSHOPPER MOUSE
Onychomys leucogaster
0 5 MM
WHITE-THROATED WOODRAT
Neotoma albigula

FIGURE 53

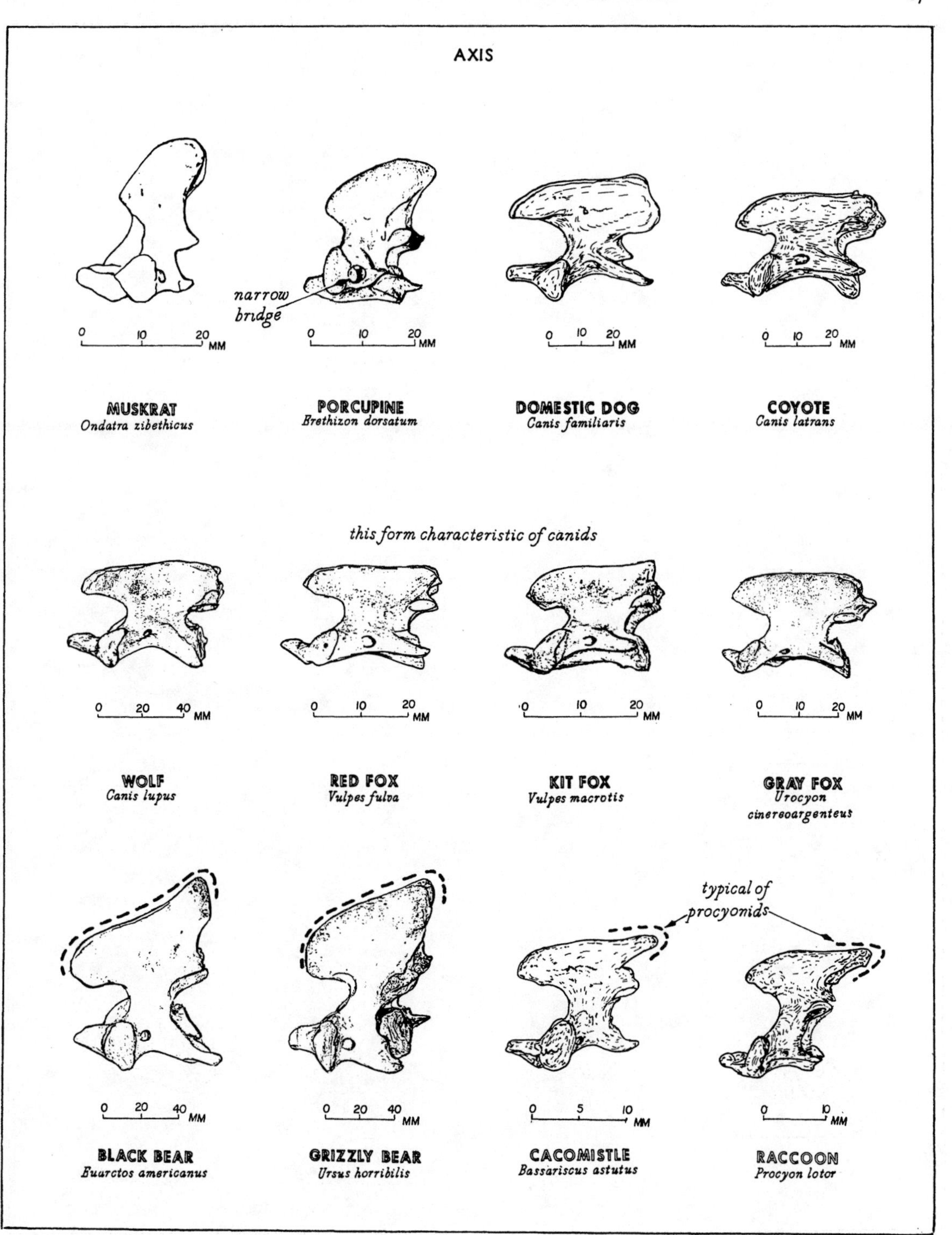
AXIS
narrow bridge
0 10 20 MM
MUSKRAT
Ondatra zibethicus
0 10 20 MM
PORCUPINE
Erethizon dorsatum
0 10 20 MM
DOMESTIC DOG
Canis familiaris
0 10 20 MM
COYOTE
Canis latrans
this form characteristic of canids
0 20 40 MM
WOLF
Canis lupus
0 10 20 MM
RED FOX
Vulpes fulva
0 10 20 MM
KIT FOX
Vulpes macrotis
0 10 20 MM
GRAY FOX
Urocyon cinereoargenteus
typical of procyonids
0 20 40 MM
BLACK BEAR
Euarctos americanus
0 20 40 MM
GRIZZLY BEAR
Ursus horribilis
0 5 10 MM
CACOMISTLE
Bassariscus astutus
0 10 MM
RACCOON
Procyon lotor

FIGURE 54

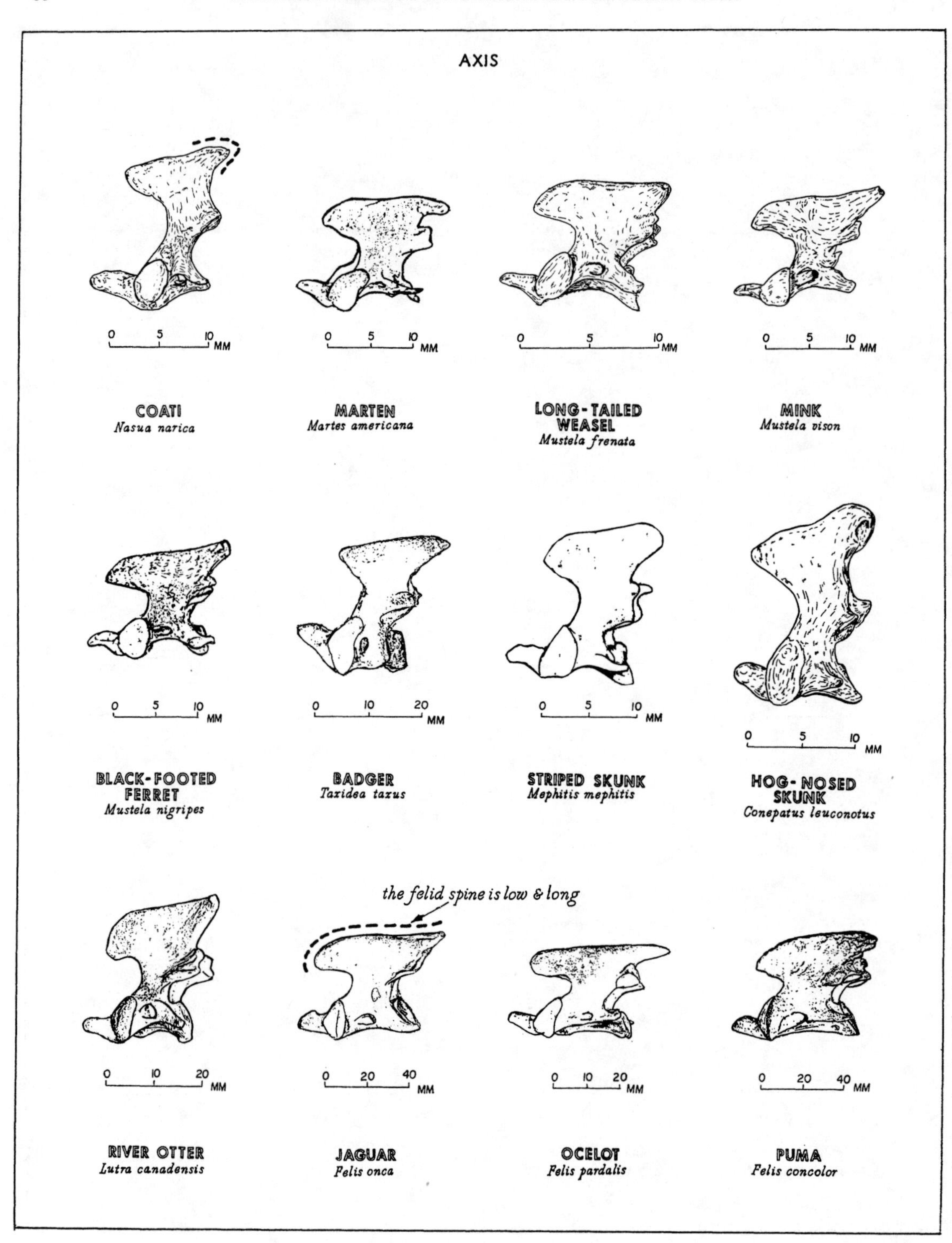
AXIS
0 5 10 MM
COATI
Nasua narica
0 5 10 MM
MARTEN
Martes americana
0 5 10 MM
LONG-TAILED WEASEL
Mustela frenata
0 5 10 MM
MINK
Mustela vison
0 5 10 MM
BLACK-FOOTED FERRET
Mustela nigripes
0 10 20 MM
BADGER
Taxidea taxus
0 5 10 MM
STRIPED SKUNK
Mephitis mephitis
0 5 10 MM
HOG-NOSED SKUNK
Conepatus leuconotus
the felid spine is low & long
0 10 20 MM
RIVER OTTER
Lutra canadensis
0 20 40 MM
JAGUAR
Felis onca
0 10 20 MM
OCELOT
Felis pardalis
0 20 40 MM
PUMA
Felis concolor

FIGURE 55

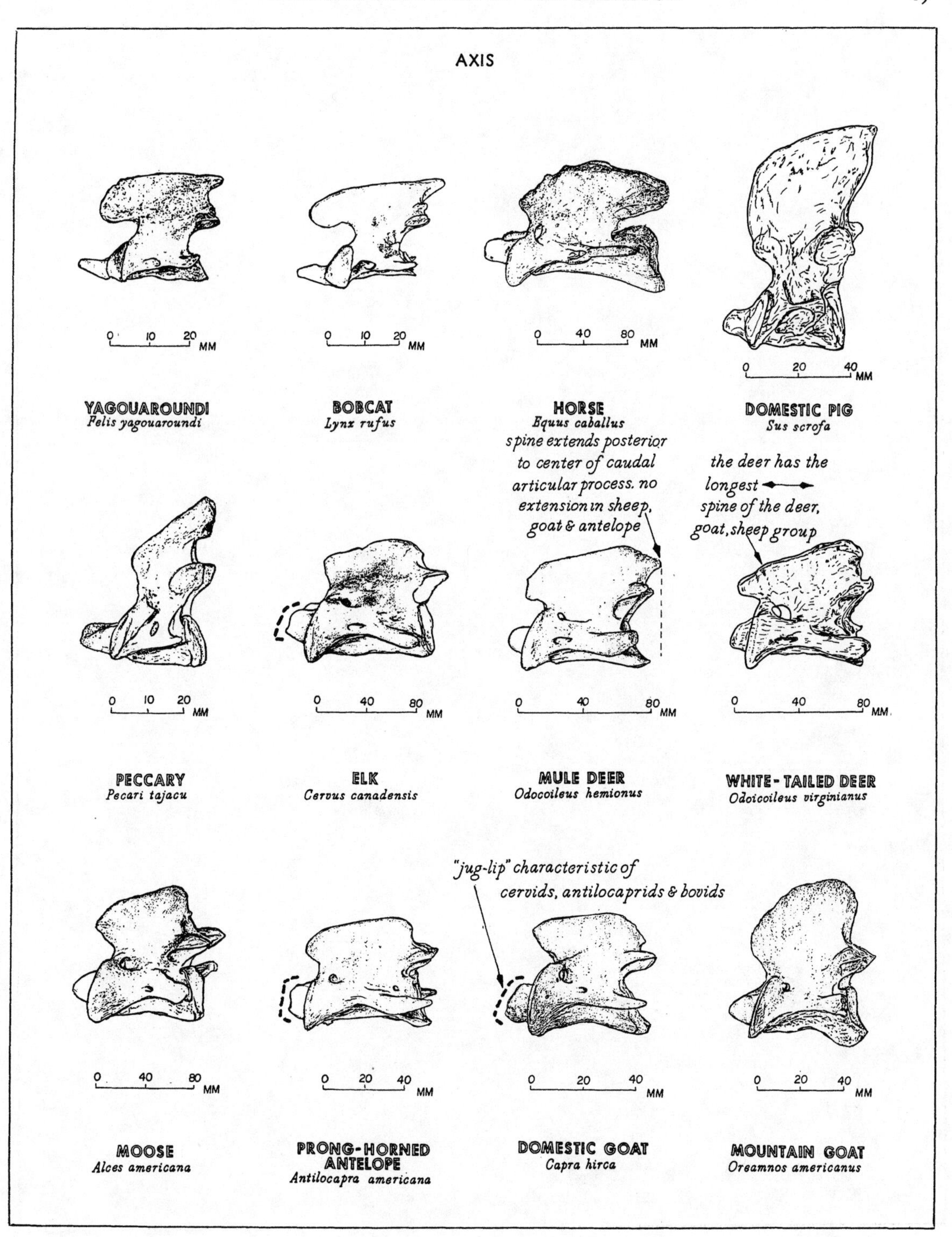
AXIS
0 10 20 MM
0 10 20 MM
0 40 80 MM
0 20 40 MM
YAGOUAROUNDI
Felis yagouaroundi
BOBCAT
Lynx rufus
HORSE
Equus caballus
DOMESTIC PIG
Sus scrofa
spine extends posterior to center of caudal articular process. no extension in sheep, goat & antelope
the deer has the longest ←→ spine of the deer, goat, sheep group
0 10 20 MM
0 40 80 MM
0 40 80 MM
0 40 80 MM
PECCARY
Pecari tajacu
ELK
Cervus canadensis
MULE DEER
Odocoileus hemionus
WHITE-TAILED DEER
Odocoileus virginianus
"jug-lip" characteristic of cervids, antilocaprids & bovids
0 40 80 MM
0 20 40 MM
0 20 40 MM
0 20 40 MM
MOOSE
Alces americana
PRONG-HORNED ANTELOPE
Antilocapra americana
DOMESTIC GOAT
Capra hirca
MOUNTAIN GOAT
Oreamnos americanus

FIGURE 56

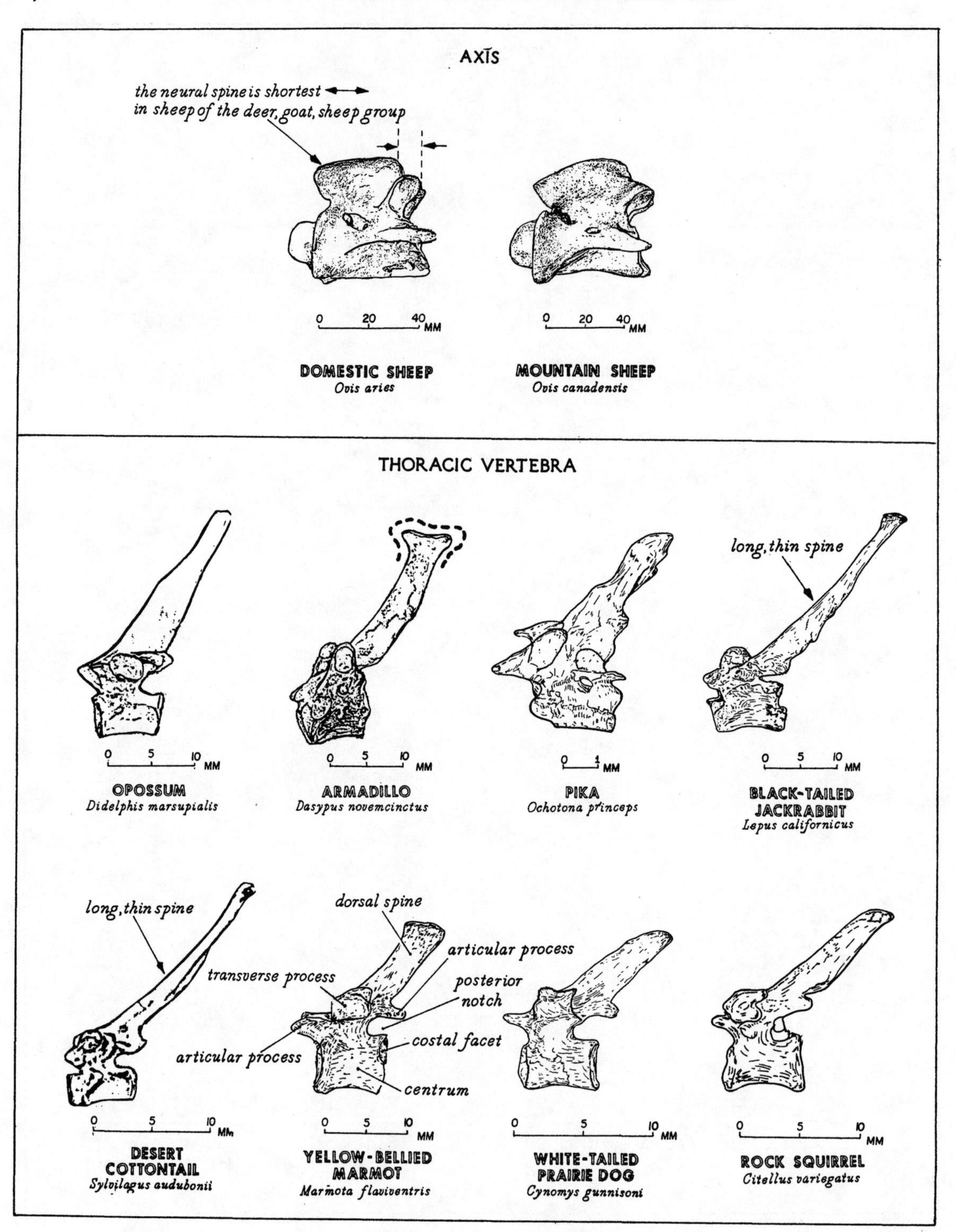
AXIS
the neural spine is shortest
in sheep of the deer, goat, sheep group
0 20 40 MM
0 20 40 MM
DOMESTIC SHEEP
Ovis aries
MOUNTAIN SHEEP
Ovis canadensis
THORACIC VERTEBRA
long, thin spine
0 5 10 MM
0 5 10 MM
0 1 MM
0 5 10 MM
OPOSSUM
Didelphis marsupialis
ARMADILLO
Dasypus novemcinctus
PIKA
Ochotona princeps
BLACK-TAILED
JACKRABBIT
Lepus californicus
long, thin spine
dorsal spine
articular process
transverse process
posterior
notch
costal facet
articular process
centrum
0 5 10 MM
0 5 10 MM
0 5 10 MM
0 5 10 MM
DESERT
COTTONTAIL
Sylvilagus audubonii
YELLOW-BELLIED
MARMOT
Marmota flaviventris
WHITE-TAILED
PRAIRIE DOG
Cynomys gunnisoni
ROCK SQUIRREL
Citellus variegatus

FIGURE 57

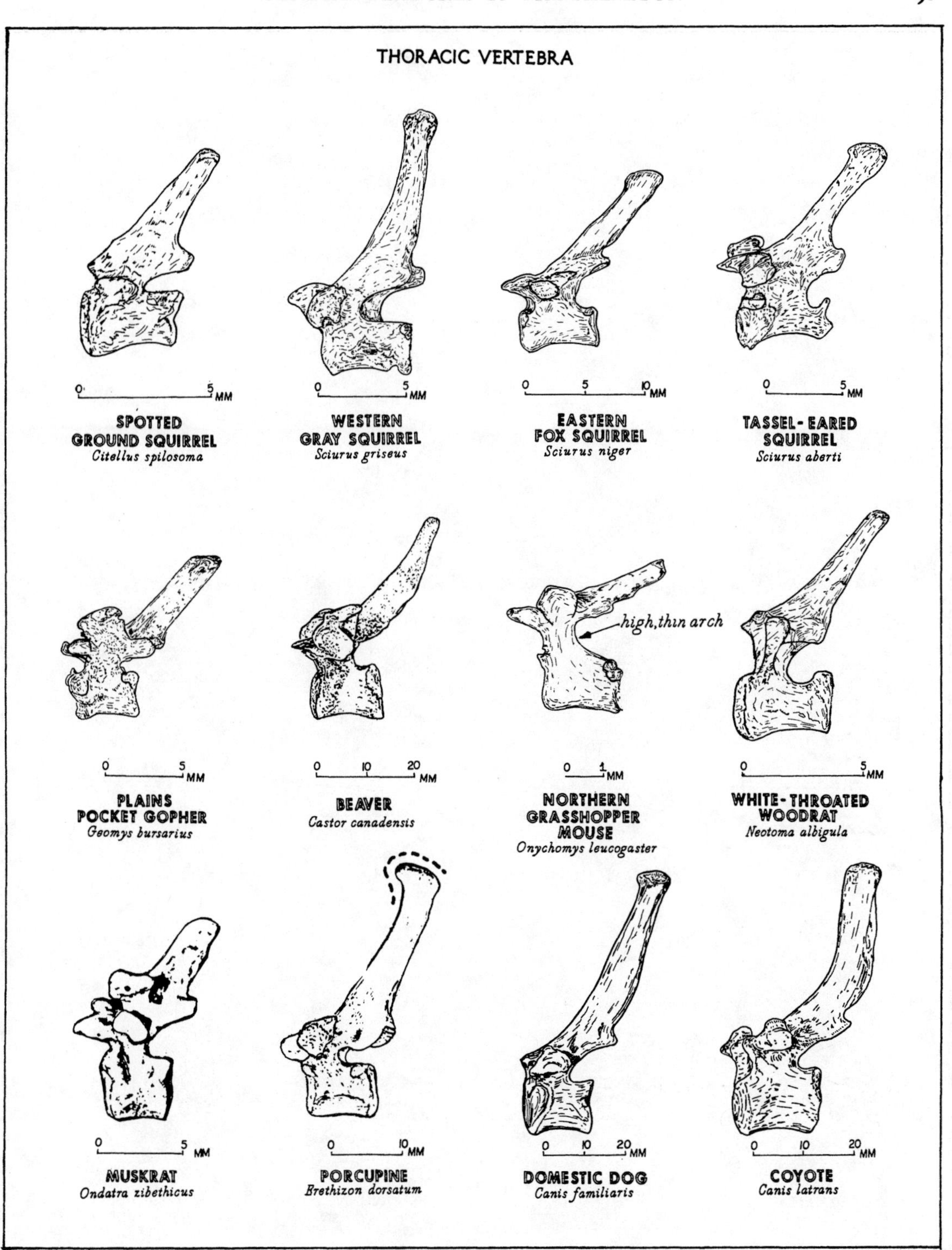
THORACIC VERTEBRA
0 5 MM
SPOTTED GROUND SQUIRREL
Citellus spilosoma
0 5 MM
WESTERN GRAY SQUIRREL
Sciurus griseus
0 5 10 MM
EASTERN FOX SQUIRREL
Sciurus niger
0 5 MM
TASSEL-EARED SQUIRREL
Sciurus aberti
0 5 MM
PLAINS POCKET GOPHER
Geomys bursarius
0 10 20 MM
BEAVER
Castor canadensis
high, thin arch
0 1 MM
NORTHERN GRASSHOPPER MOUSE
Onychomys leucogaster
0 5 MM
WHITE-THROATED WOODRAT
Neotoma albigula
0 5 MM
MUSKRAT
Ondatra zibethicus
0 10 MM
PORCUPINE
Erethizon dorsatum
0 10 20 MM
DOMESTIC DOG
Canis familiaris
0 10 20 MM
COYOTE
Canis latrans

FIGURE 58

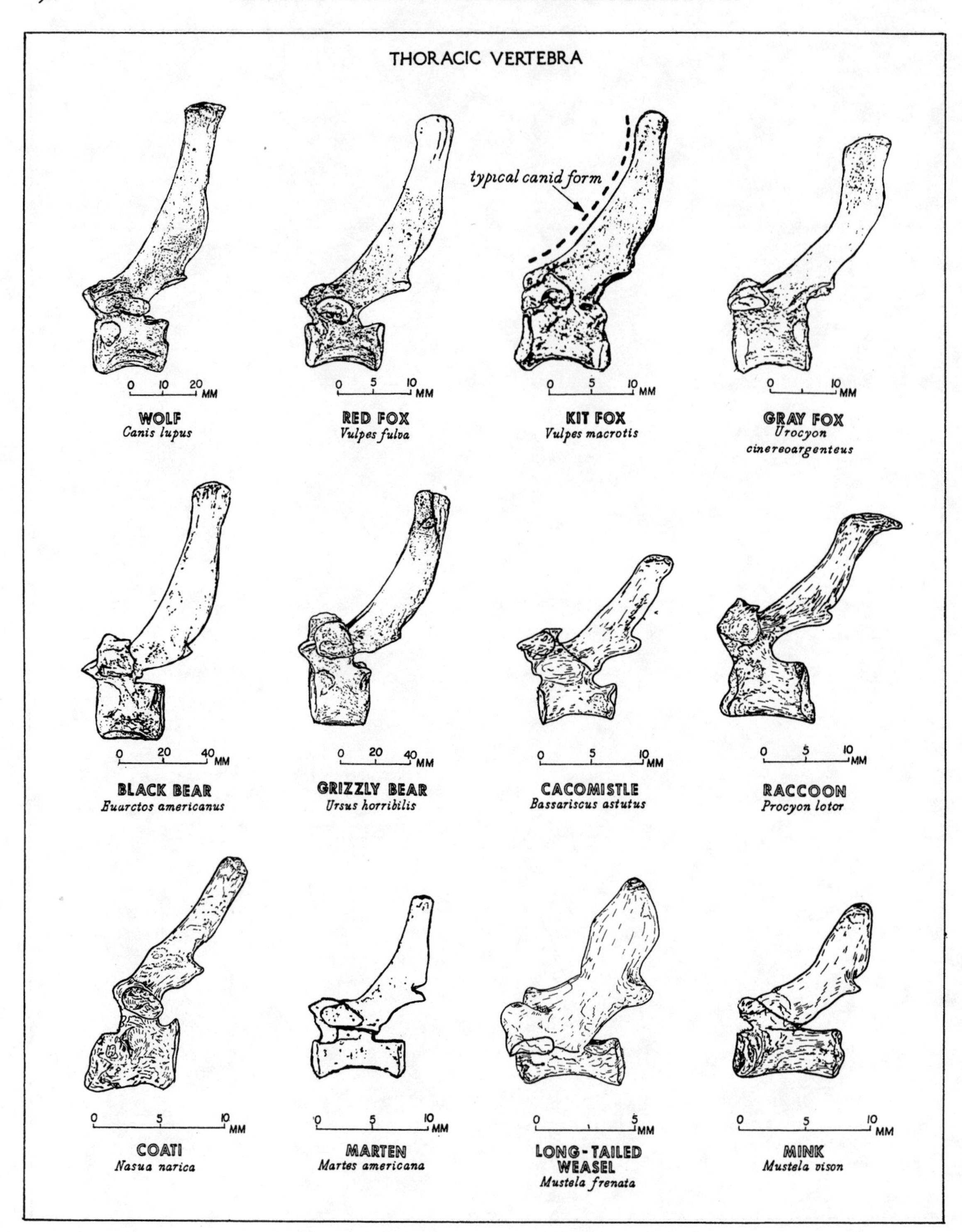
THORACIC VERTEBRA
typical canid form
0 10 20 MM
WOLF
Canis lupus
0 5 10 MM
RED FOX
Vulpes fulva
0 5 10 MM
KIT FOX
Vulpes macrotis
0 10 MM
GRAY FOX
Urocyon
cinereoargenteus
0 20 40 MM
BLACK BEAR
Euarctos americanus
0 20 40 MM
GRIZZLY BEAR
Ursus horribilis
0 5 10 MM
CACOMISTLE
Bassariscus astutus
0 5 10 MM
RACCOON
Procyon lotor
0 5 10 MM
COATI
Nasua narica
0 5 10 MM
MARTEN
Martes americana
0 5 MM
LONG-TAILED
WEASEL
Mustela frenata
0 5 10 MM
MINK
Mustela vison

FIGURE 59

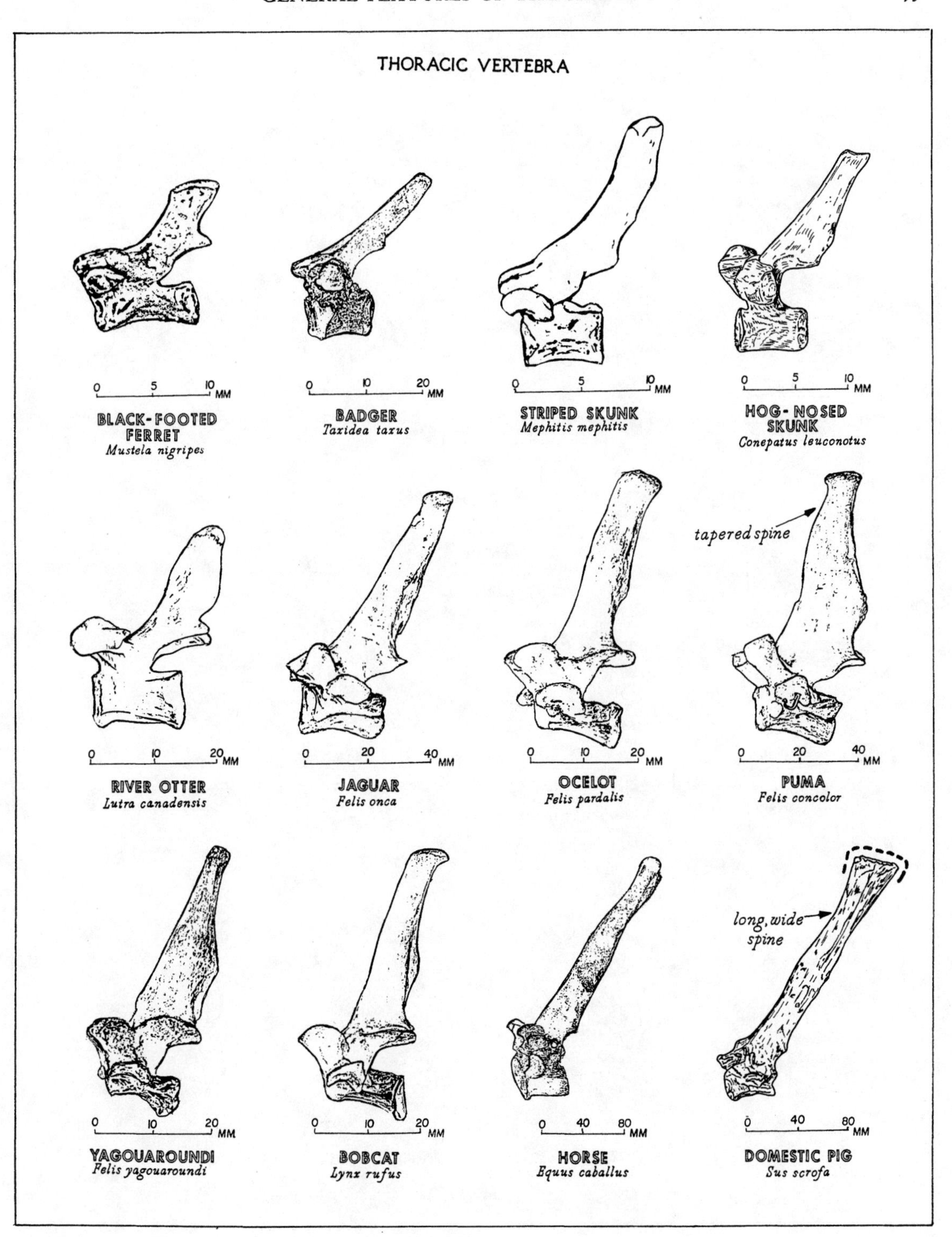
THORACIC VERTEBRA
0 5 10 MM
BLACK-FOOTED FERRET
Mustela nigripes
0 10 20 MM
BADGER
Taxidea taxus
0 5 10 MM
STRIPED SKUNK
Mephitis mephitis
0 5 10 MM
HOG-NOSED SKUNK
Conepatus leuconotus
0 10 20 MM
RIVER OTTER
Lutra canadensis
0 20 40 MM
JAGUAR
Felis onca
0 10 20 MM
OCELOT
Felis pardalis
tapered spine
0 20 40 MM
PUMA
Felis concolor
0 10 20 MM
YAGOUAROUNDI
Felis yagouaroundi
0 10 20 MM
BOBCAT
Lynx rufus
0 40 80 MM
HORSE
Equus caballus
long, wide spine
0 40 80 MM
DOMESTIC PIG
Sus scrofa

FIGURE 60

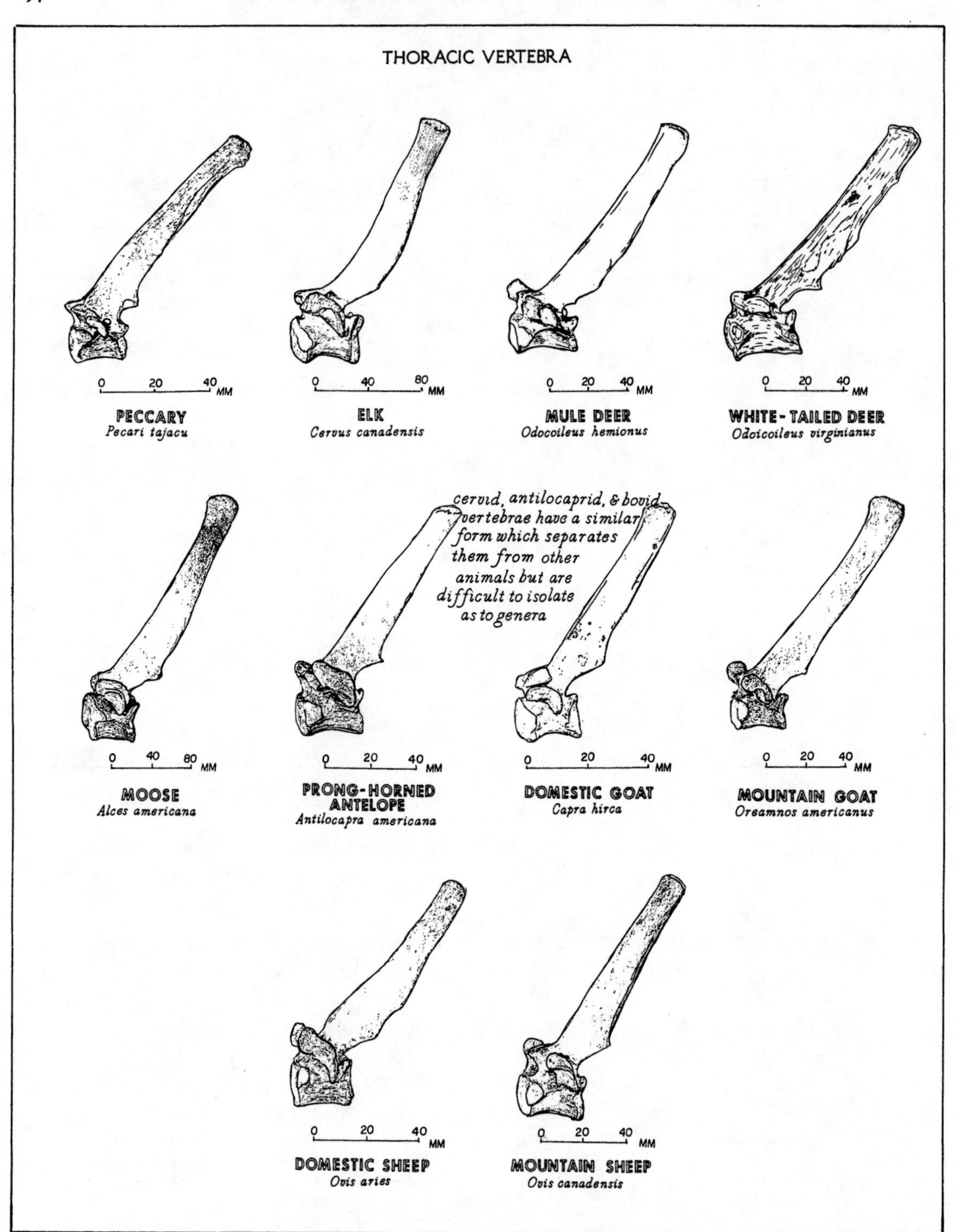
THORACIC VERTEBRA
0 20 40 MM
PECCARY
Pecari tajacu
0 40 80 MM
ELK
Cervus canadensis
0 20 40 MM
MULE DEER
Odocoileus hemionus
0 20 40 MM
WHITE-TAILED DEER
Odocoileus virginianus
cervid, antilocaprid, & bovid vertebrae have a similar form which separates them from other animals but are difficult to isolate as to genera
0 40 80 MM
MOOSE
Alces americana
0 20 40 MM
PRONG-HORNED ANTELOPE
Antilocapra americana
0 20 40 MM
DOMESTIC GOAT
Capra hirca
0 20 40 MM
MOUNTAIN GOAT
Oreamnos americanus
0 20 40 MM
DOMESTIC SHEEP
Ovis aries
0 20 40 MM
MOUNTAIN SHEEP
Ovis canadensis

FIGURE 61

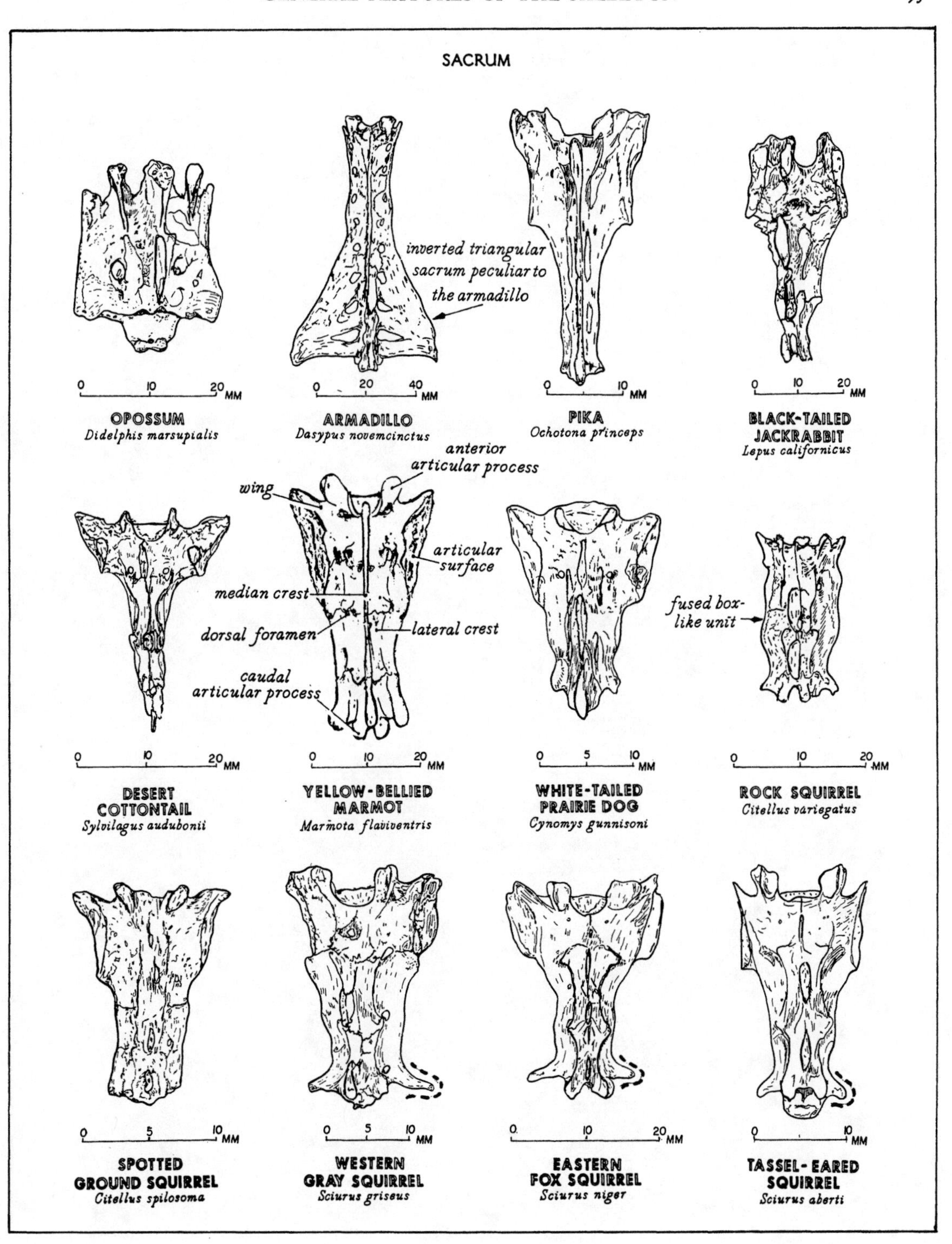
SACRUM
inverted triangular sacrum peculiar to the armadillo
0 10 20 MM
OPOSSUM
Didelphis marsupialis
0 20 40 MM
ARMADILLO
Dasypus novemcinctus
0 10 MM
PIKA
Ochotona princeps
0 10 20 MM
BLACK-TAILED JACKRABBIT
Lepus californicus
anterior articular process
wing
articular surface
median crest
dorsal foramen
lateral crest
caudal articular process
fused box-like unit
0 10 20 MM
DESERT COTTONTAIL
Sylvilagus audubonii
0 10 20 MM
YELLOW-BELLIED MARMOT
Marmota flaviventris
0 5 10 MM
WHITE-TAILED PRAIRIE DOG
Cynomys gunnisoni
0 10 20 MM
ROCK SQUIRREL
Citellus variegatus
0 5 10 MM
SPOTTED GROUND SQUIRREL
Citellus spilosoma
0 5 10 MM
WESTERN GRAY SQUIRREL
Sciurus griseus
0 10 20 MM
EASTERN FOX SQUIRREL
Sciurus niger
0 10 MM
TASSEL-EARED SQUIRREL
Sciurus aberti

FIGURE 62

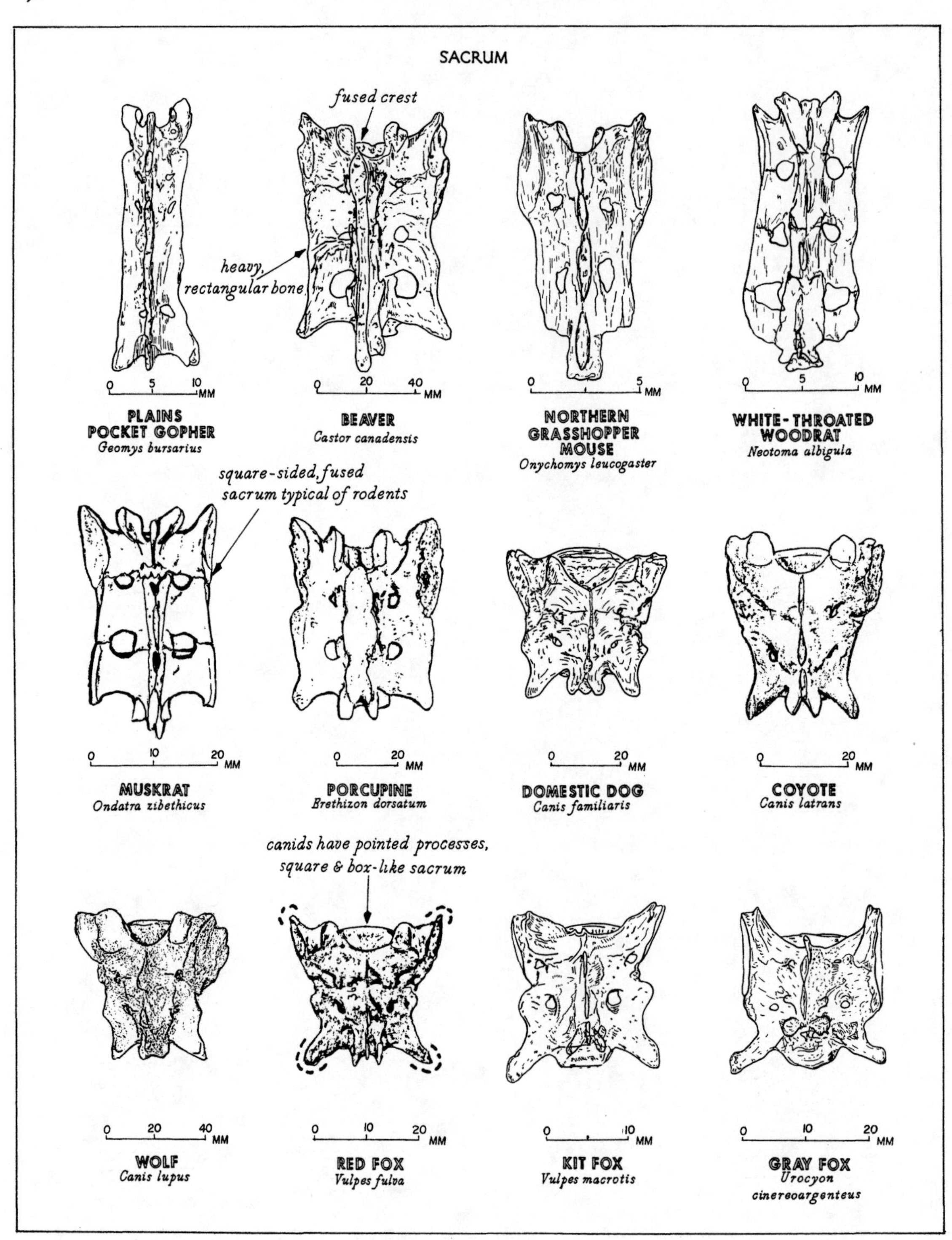
SACRUM
fused crest
heavy, rectangular bone
0 5 10 MM
PLAINS POCKET GOPHER
Geomys bursarius
0 20 40 MM
BEAVER
Castor canadensis
0 5 MM
NORTHERN GRASSHOPPER MOUSE
Onychomys leucogaster
0 5 10 MM
WHITE-THROATED WOODRAT
Neotoma albigula
square-sided, fused sacrum typical of rodents
0 10 20 MM
MUSKRAT
Ondatra zibethicus
0 20 MM
PORCUPINE
Erethizon dorsatum
0 20 MM
DOMESTIC DOG
Canis familiaris
0 20 MM
COYOTE
Canis latrans
canids have pointed processes, square & box-like sacrum
0 20 40 MM
WOLF
Canis lupus
0 10 20 MM
RED FOX
Vulpes fulva
0 10 MM
KIT FOX
Vulpes macrotis
0 10 20 MM
GRAY FOX
Urocyon cinereoargenteus

FIGURE 63

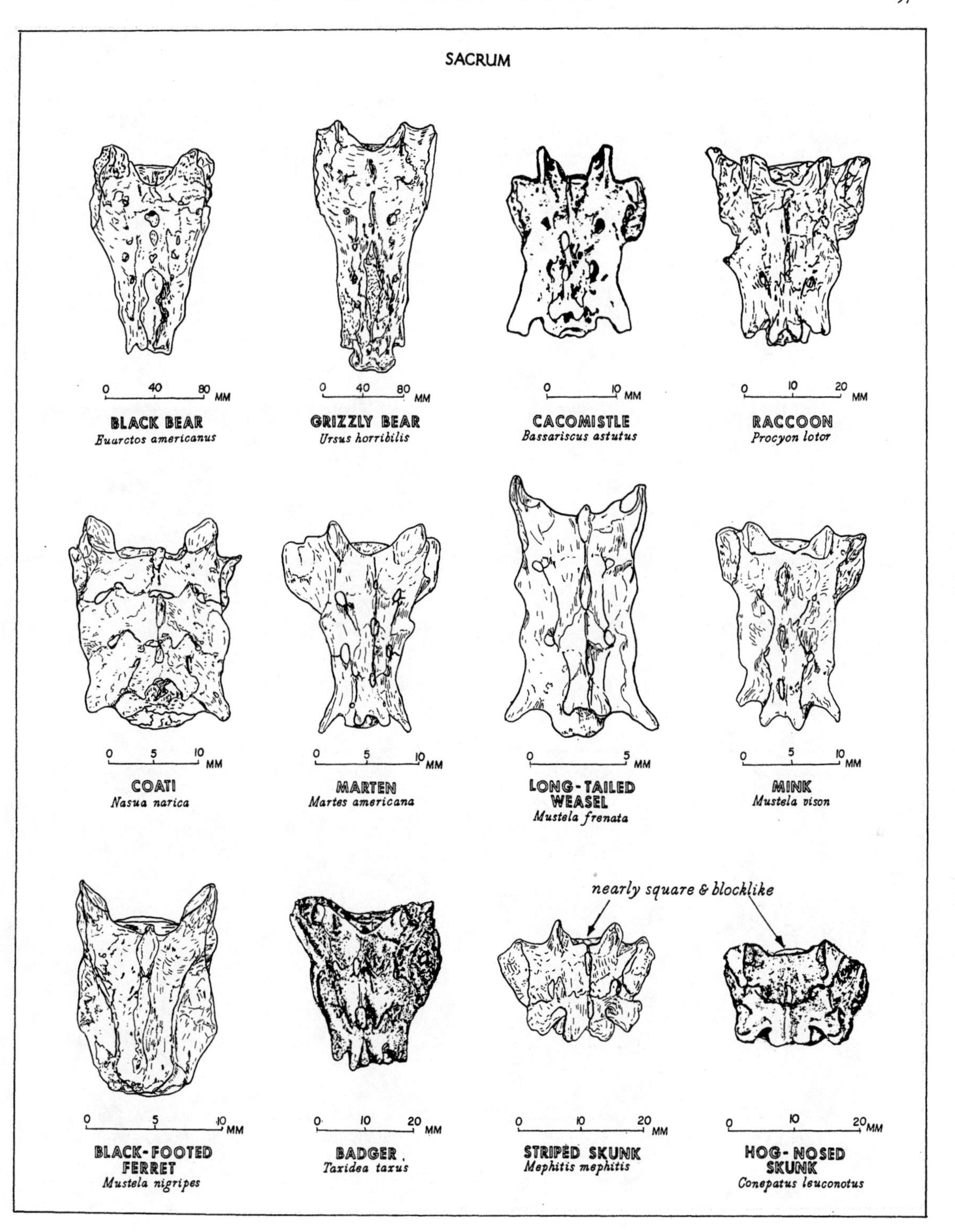
SACRUM
0 40 80 MM
BLACK BEAR
Euarctos americanus
0 40 80 MM
GRIZZLY BEAR
Ursus horribilis
0 10 MM
CACOMISTLE
Bassariscus astutus
0 10 20 MM
RACCOON
Procyon lotor
0 5 10 MM
COATI
Nasua narica
0 5 10 MM
MARTEN
Martes americana
0 5 MM
LONG-TAILED WEASEL
Mustela frenata
0 5 10 MM
MINK
Mustela vison
nearly square & blocklike
0 5 10 MM
BLACK-FOOTED FERRET
Mustela nigripes
0 10 20 MM
BADGER
Taxidea taxus
0 10 20 MM
STRIPED SKUNK
Mephitis mephitis
0 10 20 MM
HOG-NOSED SKUNK
Conepatus leuconotus

FIGURE 64

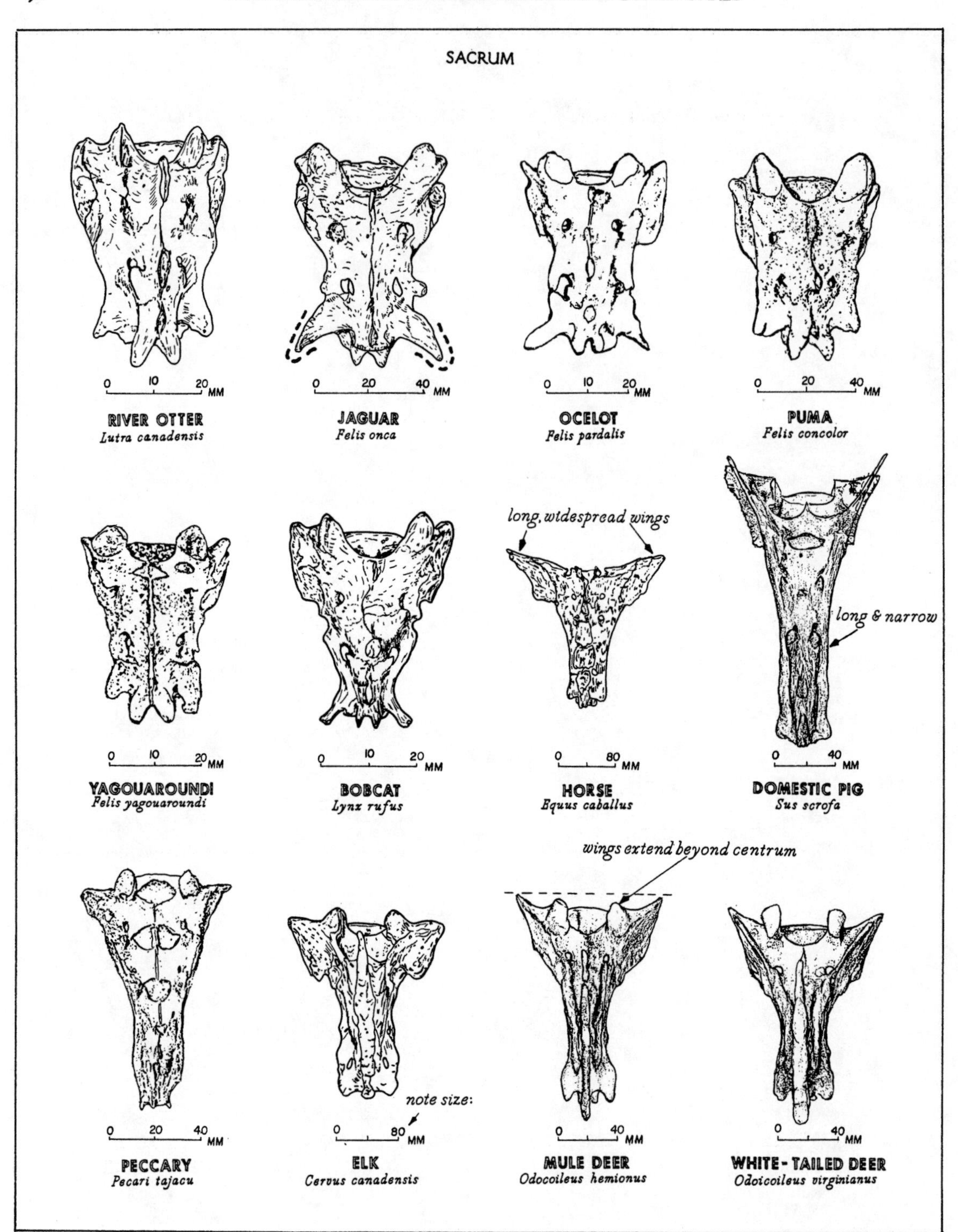
SACRUM
0 10 20 MM
RIVER OTTER
Lutra canadensis
0 20 40 MM
JAGUAR
Felis onca
0 10 20 MM
OCELOT
Felis pardalis
0 20 40 MM
PUMA
Felis concolor
0 10 20 MM
YAGOUAROUNDI
Felis yagouaroundi
0 10 20 MM
BOBCAT
Lynx rufus
long, widespread wings
0 80 MM
HORSE
Equus caballus
long & narrow
0 40 MM
DOMESTIC PIG
Sus scrofa
wings extend beyond centrum
0 20 40 MM
PECCARY
Pecari tajacu
note size:
0 80 MM
ELK
Cervus canadensis
0 40 MM
MULE DEER
Odocoileus hemionus
0 40 MM
WHITE-TAILED DEER
Odocoileus virginianus

FIGURE 65

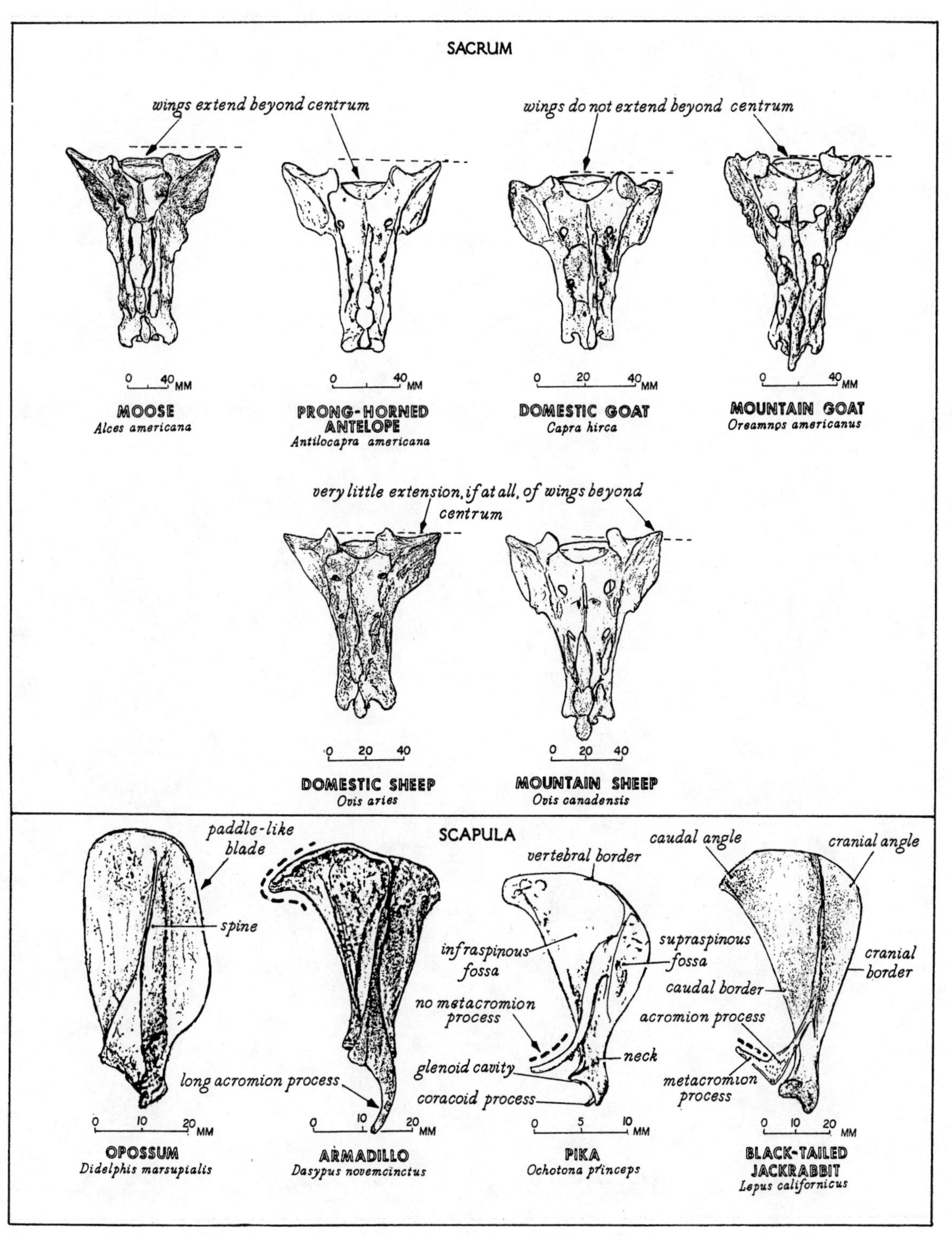
SACRUM
wings extend beyond centrum
wings do not extend beyond centrum
0 40 MM
MOOSE
Alces americana
0 40 MM
PRONG-HORNED ANTELOPE
Antilocapra americana
0 20 40 MM
DOMESTIC GOAT
Capra hirca
0 40 MM
MOUNTAIN GOAT
Oreamnos americanus
very little extension, if at all, of wings beyond centrum
0 20 40
DOMESTIC SHEEP
Ovis aries
0 20 40
MOUNTAIN SHEEP
Ovis canadensis
SCAPULA
paddle-like blade
spine
long acromion process
0 10 20 MM
OPOSSUM
Didelphis marsupialis
0 10 20 MM
ARMADILLO
Dasypus novemcinctus
vertebral border
infraspinous fossa
supraspinous fossa
no metacromion process
glenoid cavity
neck
coracoid process
0 5 10 MM
PIKA
Ochotona princeps
caudal angle
cranial angle
cranial border
caudal border
acromion process
metacromion process
0 10 20 MM
BLACK-TAILED JACKRABBIT
Lepus californicus

FIGURE 66

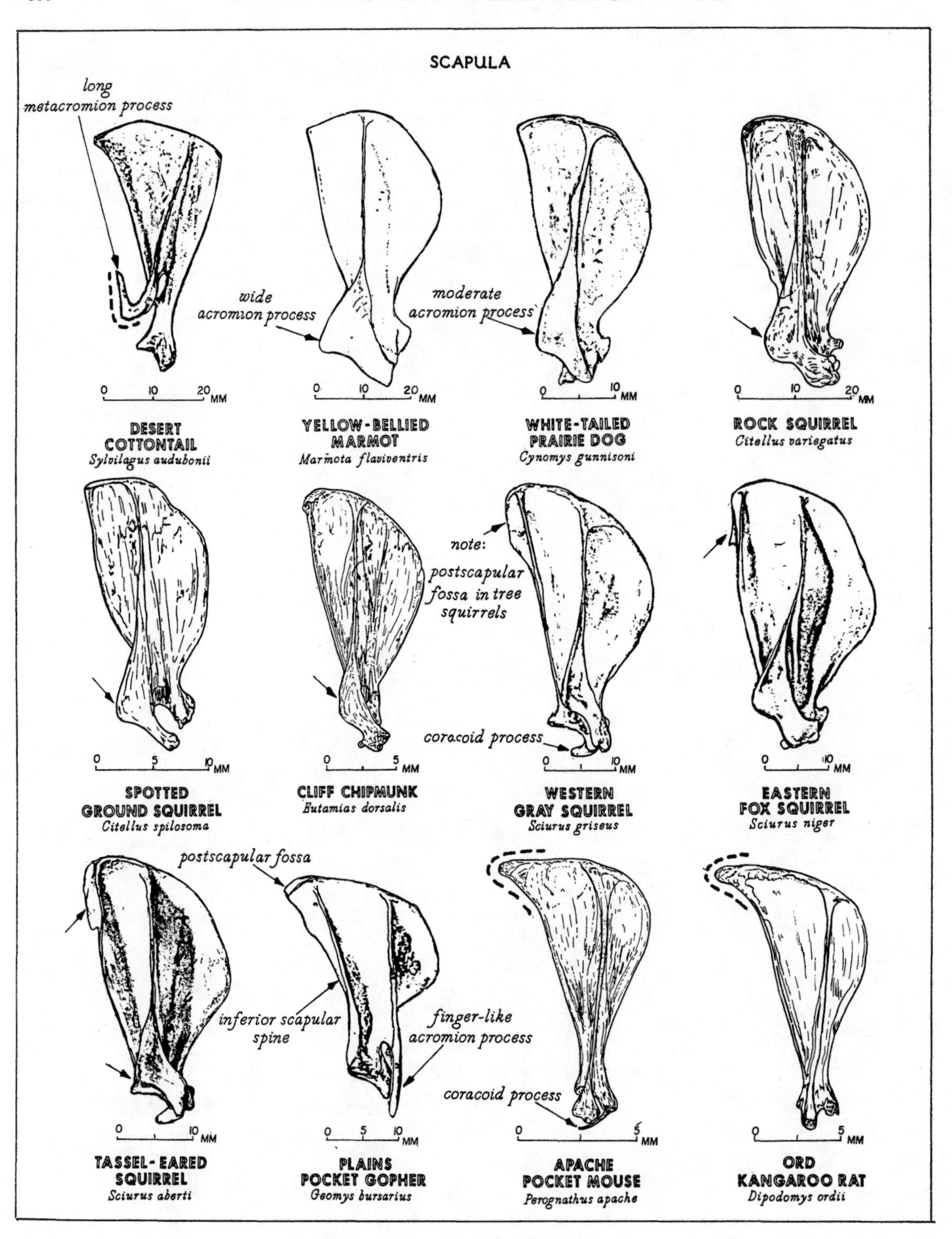
SCAPULA
long
metacromion process
wide
acromion process
moderate
acromion process
0 10 20 MM
DESERT
COTTONTAIL
Sylvilagus audubonii
YELLOW-BELLIED
MARMOT
Marmota flaviventris
WHITE-TAILED
PRAIRIE DOG
Cynomys gunnisoni
ROCK SQUIRREL
Citellus variegatus
note:
postscapular
fossa in tree
squirrels
coracoid process
SPOTTED
GROUND SQUIRREL
Citellus spilosoma
CLIFF CHIPMUNK
Eutamias dorsalis
WESTERN
GRAY SQUIRREL
Sciurus griseus
EASTERN
FOX SQUIRREL
Sciurus niger
postscapular fossa
inferior scapular
spine
finger-like
acromion process
coracoid process
TASSEL-EARED
SQUIRREL
Sciurus aberti
PLAINS
POCKET GOPHER
Geomys bursarius
APACHE
POCKET MOUSE
Perognathus apache
ORD
KANGAROO RAT
Dipodomys ordii

FIGURE 67

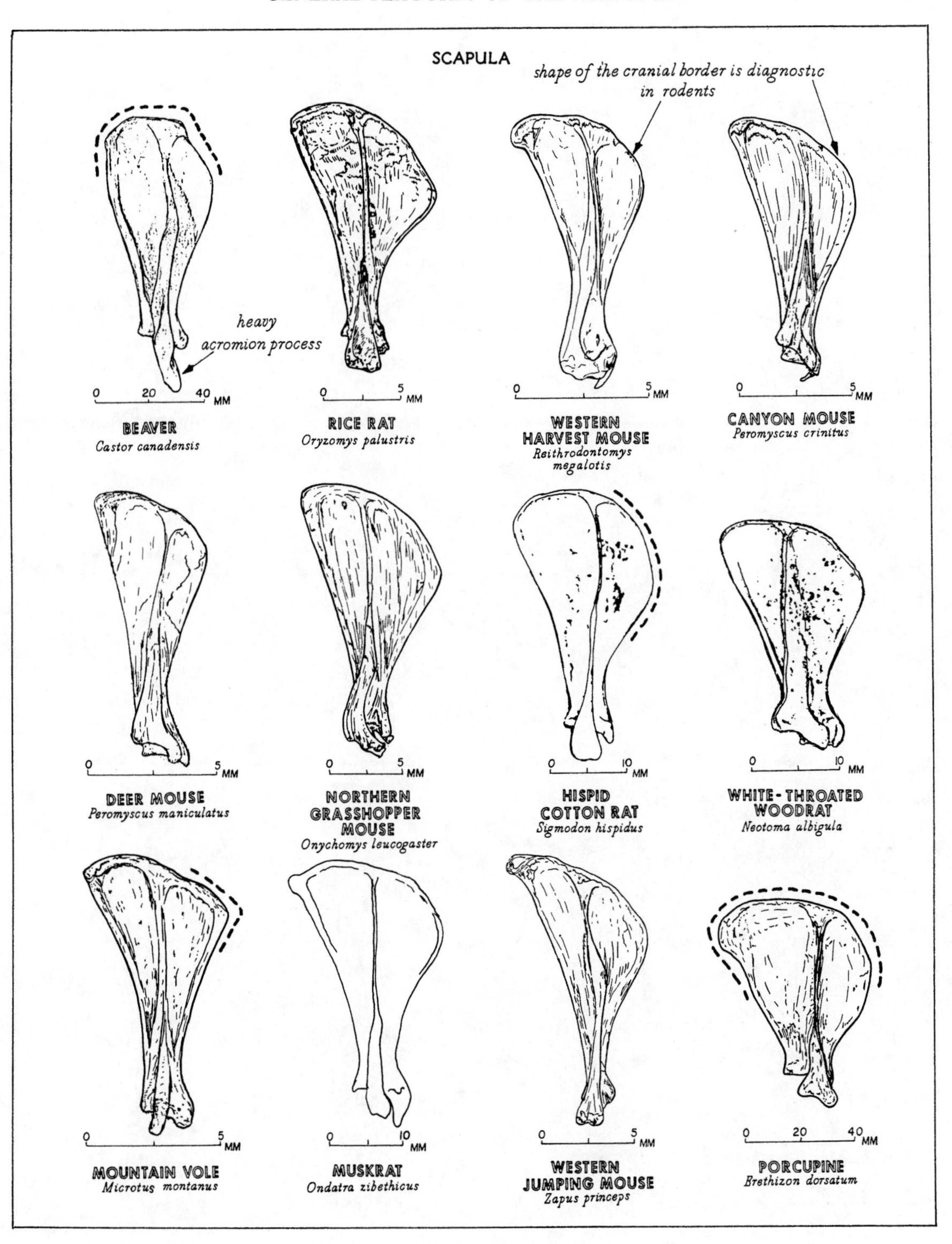
SCAPULA
shape of the cranial border is diagnostic in rodents
heavy acromion process
0 20 40 MM
BEAVER
Castor canadensis
0 5 MM
RICE RAT
Oryzomys palustris
0 5 MM
WESTERN HARVEST MOUSE
Reithrodontomys megalotis
0 5 MM
CANYON MOUSE
Peromyscus crinitus
0 5 MM
DEER MOUSE
Peromyscus maniculatus
0 5 MM
NORTHERN GRASSHOPPER MOUSE
Onychomys leucogaster
0 10 MM
HISPID COTTON RAT
Sigmodon hispidus
0 10 MM
WHITE-THROATED WOODRAT
Neotoma albigula
0 5 MM
MOUNTAIN VOLE
Microtus montanus
0 10 MM
MUSKRAT
Ondatra zibethicus
0 5 MM
WESTERN JUMPING MOUSE
Zapus princeps
0 20 40 MM
PORCUPINE
Erethizon dorsatum

FIGURE 68

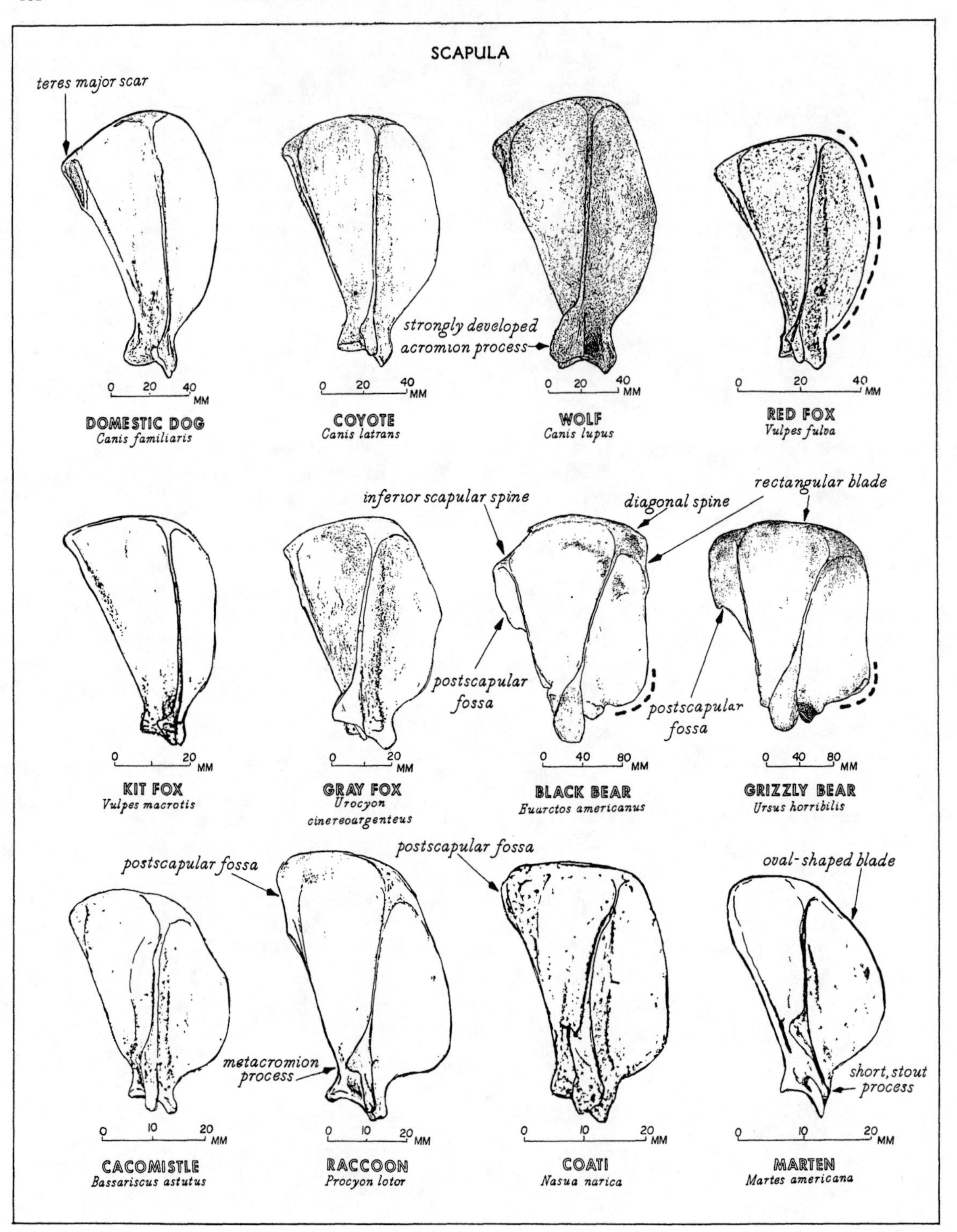
SCAPULA
teres major scar
strongly developed acromion process
0 20 40 MM
DOMESTIC DOG
Canis familiaris
COYOTE
Canis latrans
WOLF
Canis lupus
RED FOX
Vulpes fulva
inferior scapular spine
diagonal spine
rectangular blade
postscapular fossa
0 20 MM
0 40 80 MM
KIT FOX
Vulpes macrotis
GRAY FOX
Urocyon cinereoargenteus
BLACK BEAR
Euarctos americanus
GRIZZLY BEAR
Ursus horribilis
oval-shaped blade
metacromion process
short, stout process
0 10 20 MM
CACOMISTLE
Bassariscus astutus
RACCOON
Procyon lotor
COATI
Nasua narica
MARTEN
Martes americana

FIGURE 69

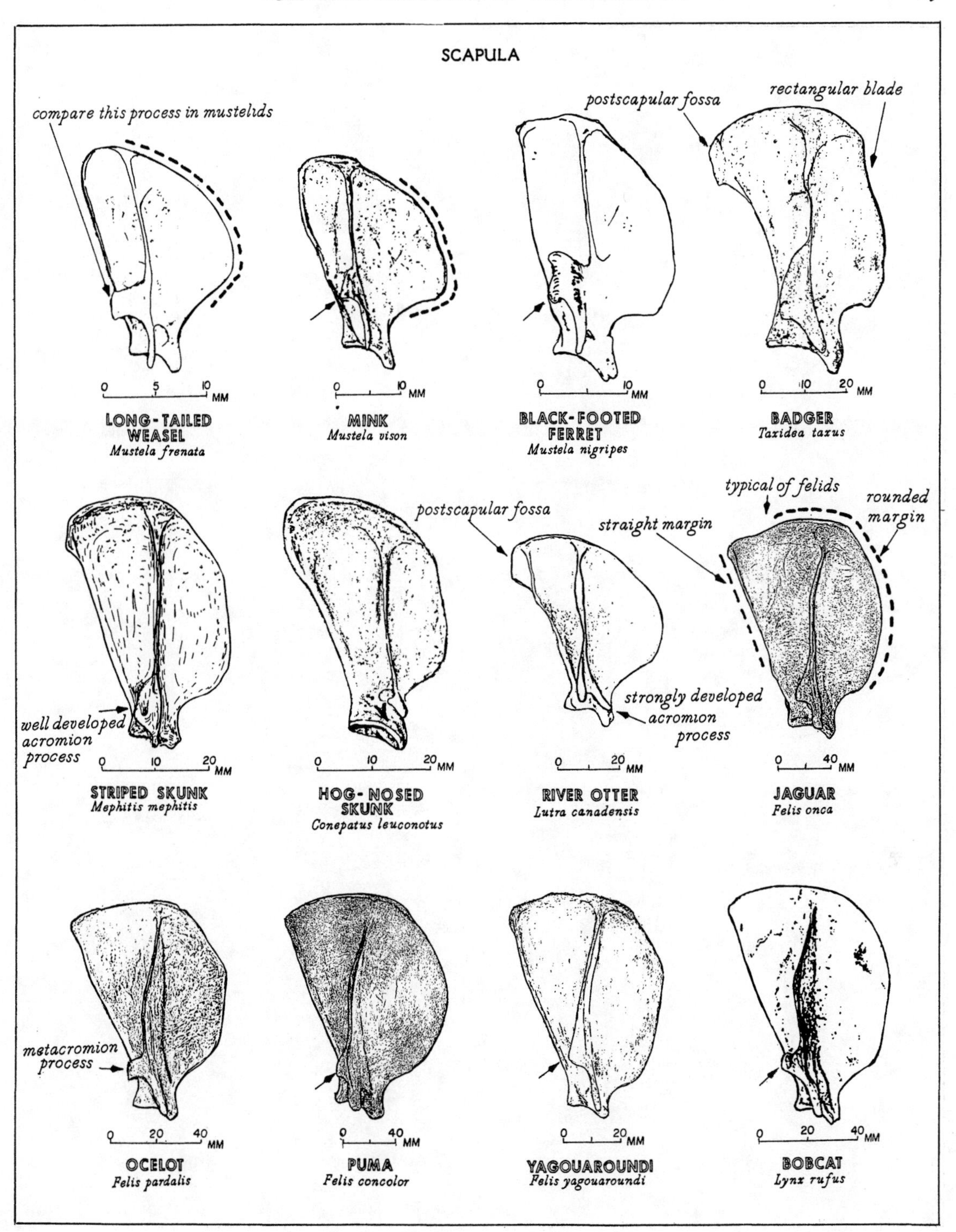
SCAPULA
compare this process in mustelids
postscapular fossa
rectangular blade
0 5 10 MM
LONG-TAILED WEASEL
Mustela frenata
0 10 MM
MINK
Mustela vison
0 10 MM
BLACK-FOOTED FERRET
Mustela nigripes
0 10 20 MM
BADGER
Taxidea taxus
typical of felids
rounded margin
postscapular fossa
straight margin
well developed acromion process
strongly developed acromion process
0 10 20 MM
STRIPED SKUNK
Mephitis mephitis
0 10 20 MM
HOG-NOSED SKUNK
Conepatus leuconotus
0 20 MM
RIVER OTTER
Lutra canadensis
0 40 MM
JAGUAR
Felis onca
metacromion process
0 20 40 MM
OCELOT
Felis pardalis
0 40 MM
PUMA
Felis concolor
0 20 MM
YAGOUAROUNDI
Felis yagouaroundi
0 20 40 MM
BOBCAT
Lynx rufus

FIGURE 70

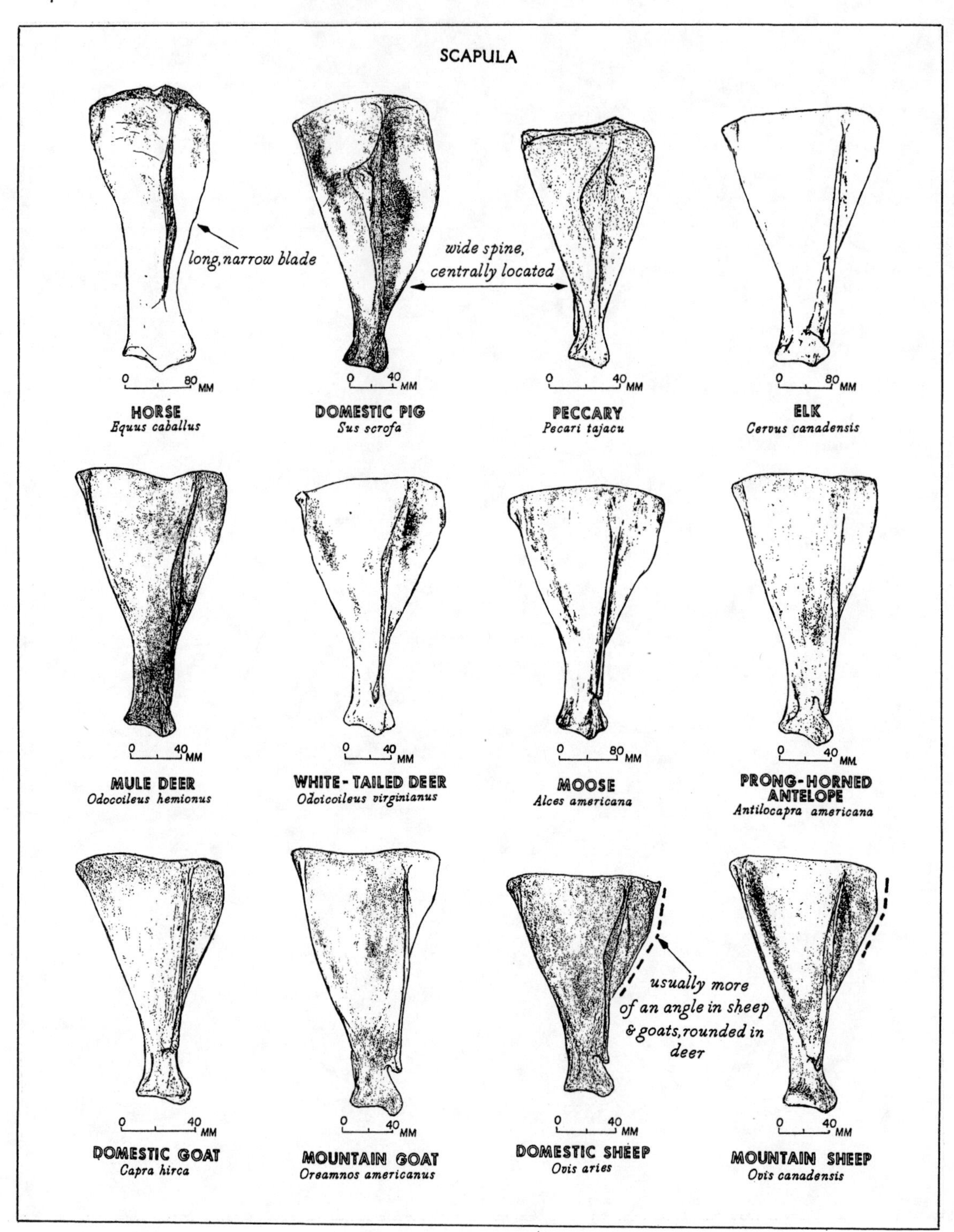
SCAPULA
long, narrow blade
wide spine, centrally located
0 80 MM
HORSE
Equus caballus
0 40 MM
DOMESTIC PIG
Sus scrofa
0 40 MM
PECCARY
Pecari tajacu
0 80 MM
ELK
Cervus canadensis
0 40 MM
MULE DEER
Odocoileus hemionus
0 40 MM
WHITE-TAILED DEER
Odoicoileus virginianus
0 80 MM
MOOSE
Alces americana
0 40 MM
PRONG-HORNED ANTELOPE
Antilocapra americana
usually more of an angle in sheep & goats, rounded in deer
0 40 MM
DOMESTIC GOAT
Capra hirca
0 40 MM
MOUNTAIN GOAT
Oreamnos americanus
0 40 MM
DOMESTIC SHEEP
Ovis aries
0 40 MM
MOUNTAIN SHEEP
Ovis canadensis

FIGURE 71

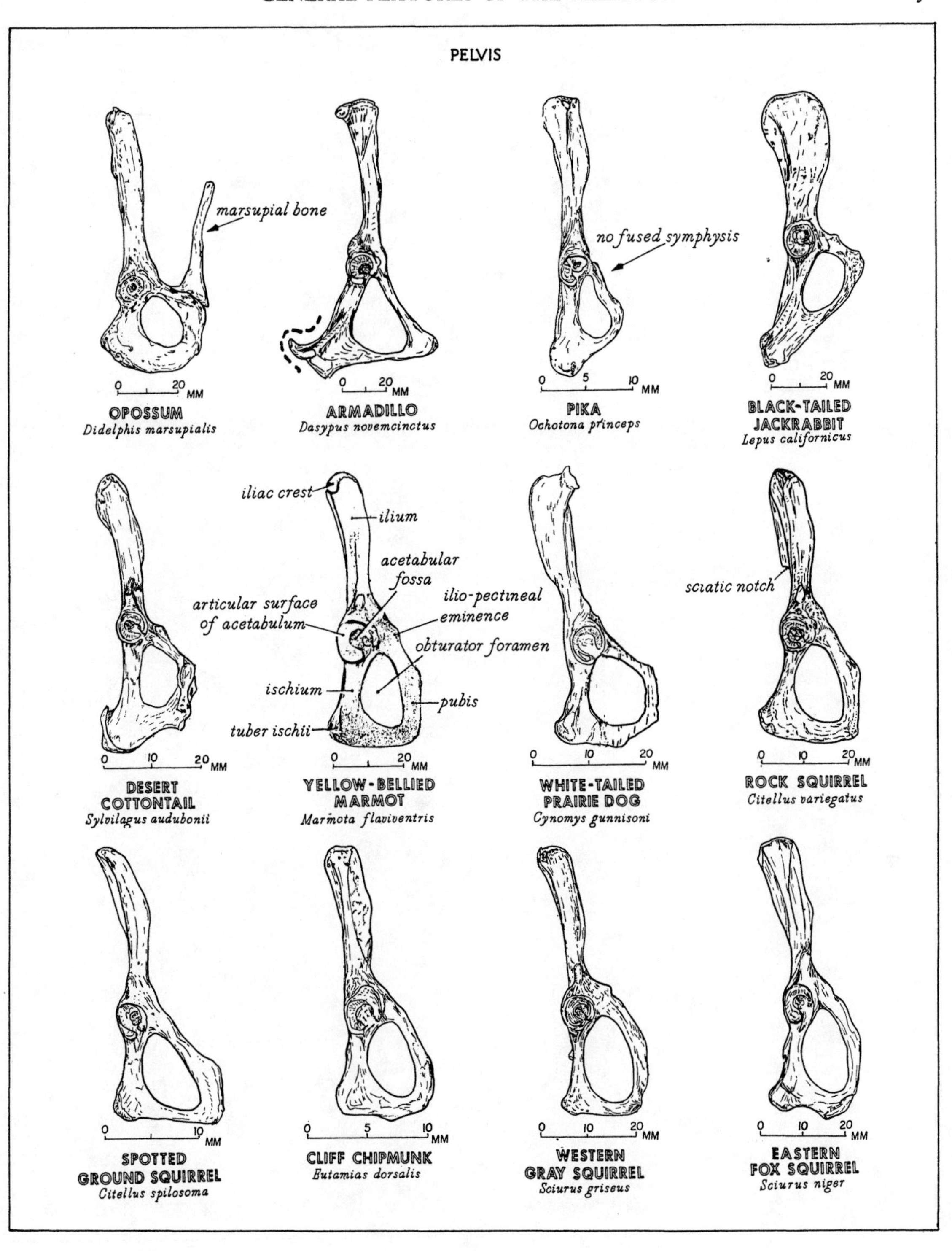
PELVIS
marsupial bone
0 20 MM
OPOSSUM
Didelphis marsupialis
0 20 MM
ARMADILLO
Dasypus novemcinctus
no fused symphysis
0 5 10 MM
PIKA
Ochotona princeps
0 20 MM
BLACK-TAILED JACKRABBIT
Lepus californicus
0 10 20 MM
DESERT COTTONTAIL
Sylvilagus audubonii
iliac crest
ilium
acetabular fossa
articular surface of acetabulum
ilio-pectineal eminence
obturator foramen
ischium
pubis
tuber ischii
0 20 MM
YELLOW-BELLIED MARMOT
Marmota flaviventris
0 10 20 MM
WHITE-TAILED PRAIRIE DOG
Cynomys gunnisoni
sciatic notch
0 10 20 MM
ROCK SQUIRREL
Citellus variegatus
0 10 MM
SPOTTED GROUND SQUIRREL
Citellus spilosoma
0 5 10 MM
CLIFF CHIPMUNK
Eutamias dorsalis
0 10 20 MM
WESTERN GRAY SQUIRREL
Sciurus griseus
0 10 20 MM
EASTERN FOX SQUIRREL
Sciurus niger

FIGURE 72

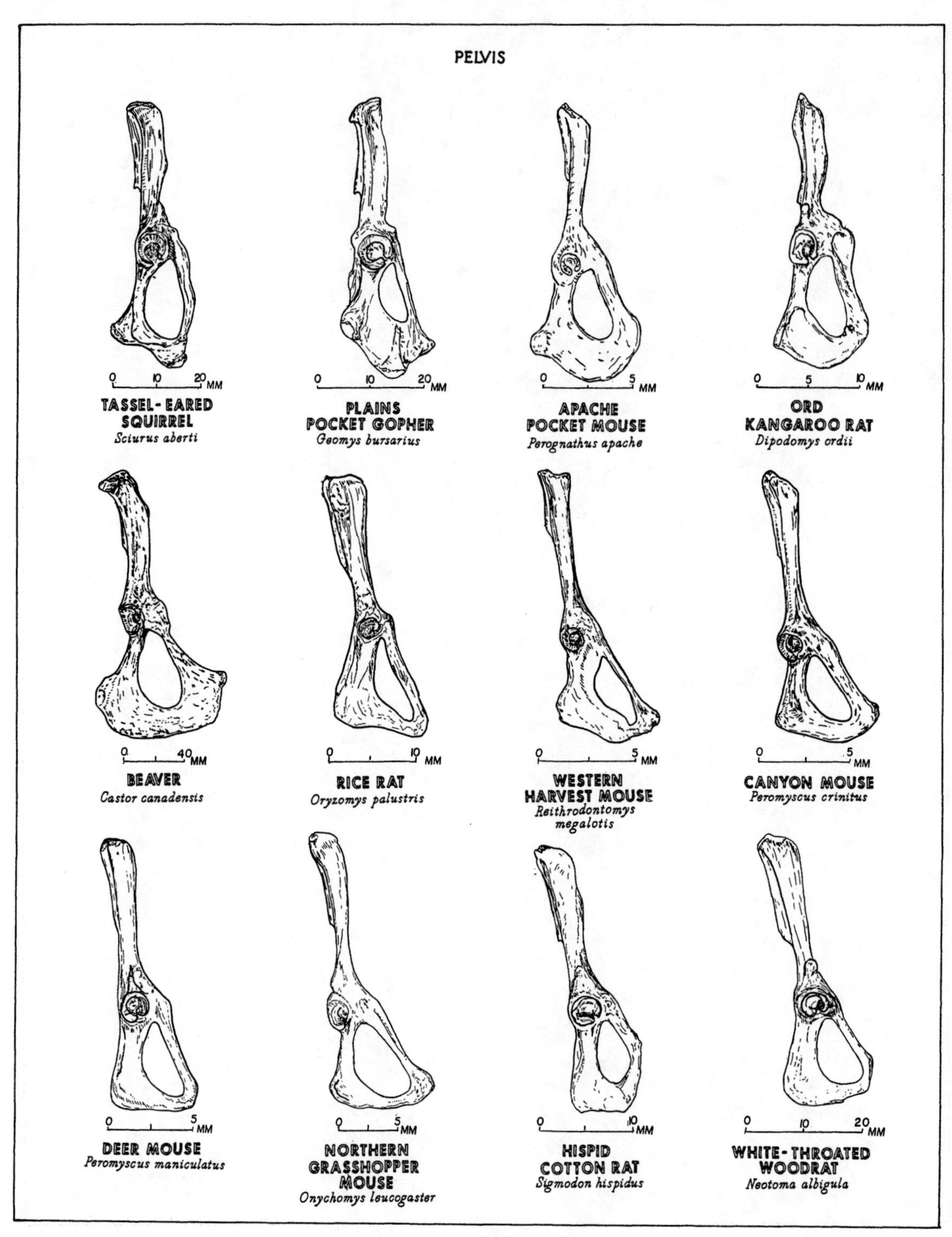
PELVIS
0 10 20 MM
TASSEL-EARED SQUIRREL
Sciurus aberti
0 10 20 MM
PLAINS POCKET GOPHER
Geomys bursarius
0 5 MM
APACHE POCKET MOUSE
Perognathus apache
0 5 10 MM
ORD KANGAROO RAT
Dipodomys ordii
0 40 MM
BEAVER
Castor canadensis
0 10 MM
RICE RAT
Oryzomys palustris
0 5 MM
WESTERN HARVEST MOUSE
Reithrodontomys megalotis
0 5 MM
CANYON MOUSE
Peromyscus crinitus
0 5 MM
DEER MOUSE
Peromyscus maniculatus
0 5 MM
NORTHERN GRASSHOPPER MOUSE
Onychomys leucogaster
0 10 MM
HISPID COTTON RAT
Sigmodon hispidus
0 10 20 MM
WHITE-THROATED WOODRAT
Neotoma albigula

FIGURE 73

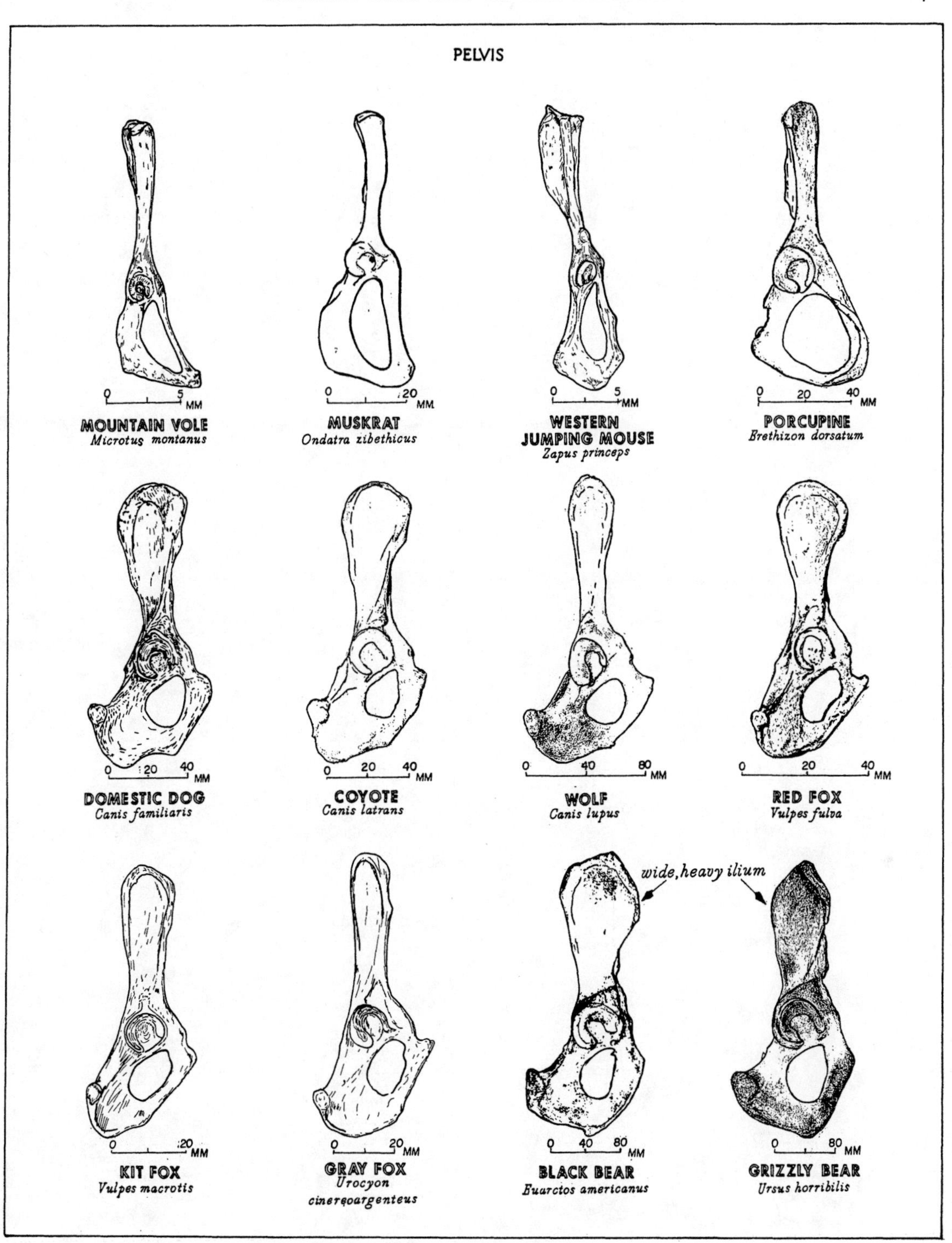
PELVIS
0 5 MM
MOUNTAIN VOLE
Microtus montanus
0 20 MM
MUSKRAT
Ondatra zibethicus
0 5 MM
WESTERN JUMPING MOUSE
Zapus princeps
0 20 40 MM
PORCUPINE
Brethizon dorsatum
0 20 40 MM
DOMESTIC DOG
Canis familiaris
0 20 40 MM
COYOTE
Canis latrans
0 40 80 MM
WOLF
Canis lupus
0 20 40 MM
RED FOX
Vulpes fulva
wide, heavy ilium
0 20 MM
KIT FOX
Vulpes macrotis
0 20 MM
GRAY FOX
Urocyon cinereoargenteus
0 40 80 MM
BLACK BEAR
Euarctos americanus
0 80 MM
GRIZZLY BEAR
Ursus horribilis

FIGURE 74

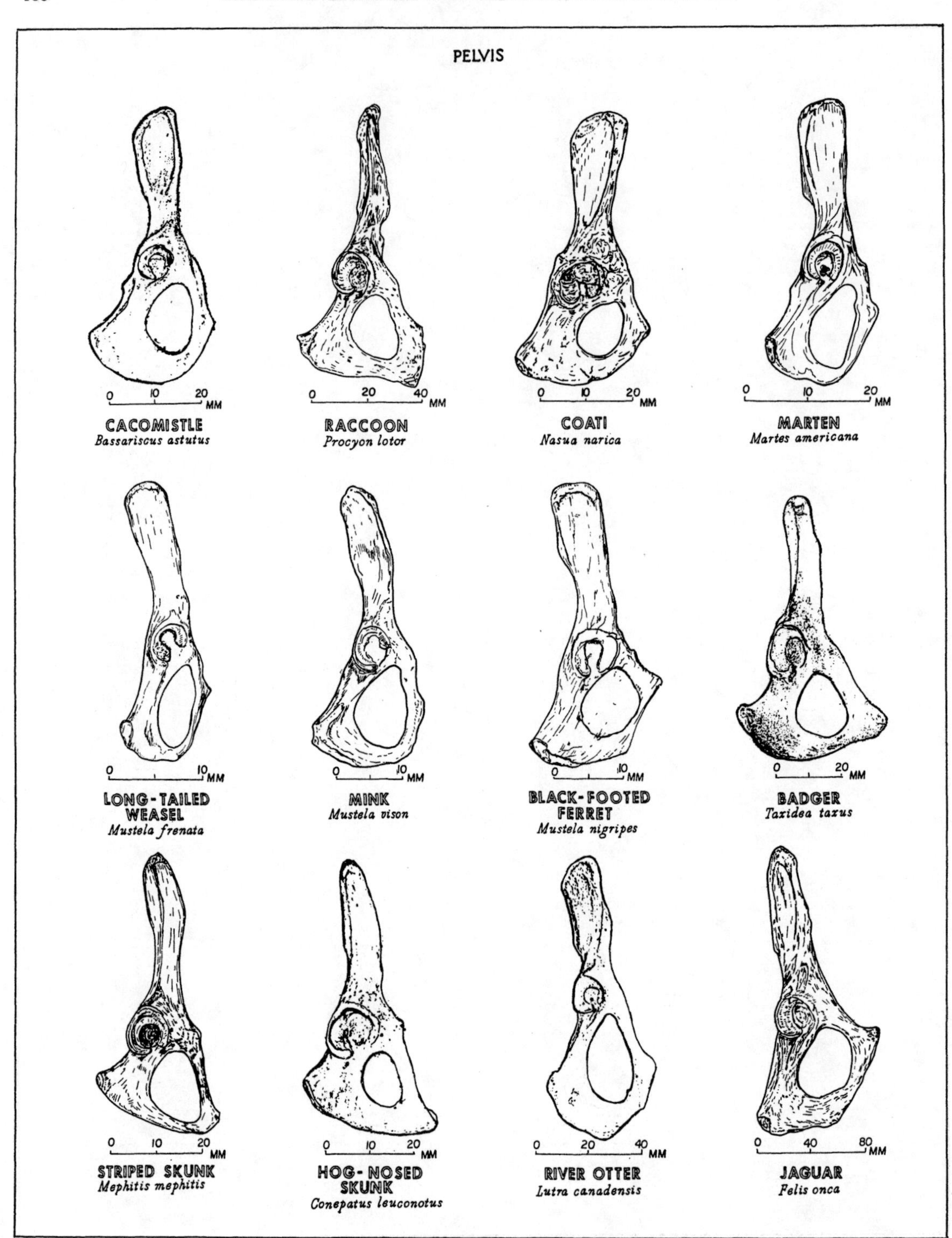
PELVIS
0 10 20 MM
CACOMISTLE
Bassariscus astutus
0 20 40 MM
RACCOON
Procyon lotor
0 10 20 MM
COATI
Nasua narica
0 10 20 MM
MARTEN
Martes americana
0 10 MM
LONG-TAILED WEASEL
Mustela frenata
0 10 MM
MINK
Mustela vison
0 10 MM
BLACK-FOOTED FERRET
Mustela nigripes
0 20 MM
BADGER
Taxidea taxus
0 10 20 MM
STRIPED SKUNK
Mephitis mephitis
0 10 20 MM
HOG-NOSED SKUNK
Conepatus leuconotus
0 20 40 MM
RIVER OTTER
Lutra canadensis
0 40 80 MM
JAGUAR
Felis onca

FIGURE 75

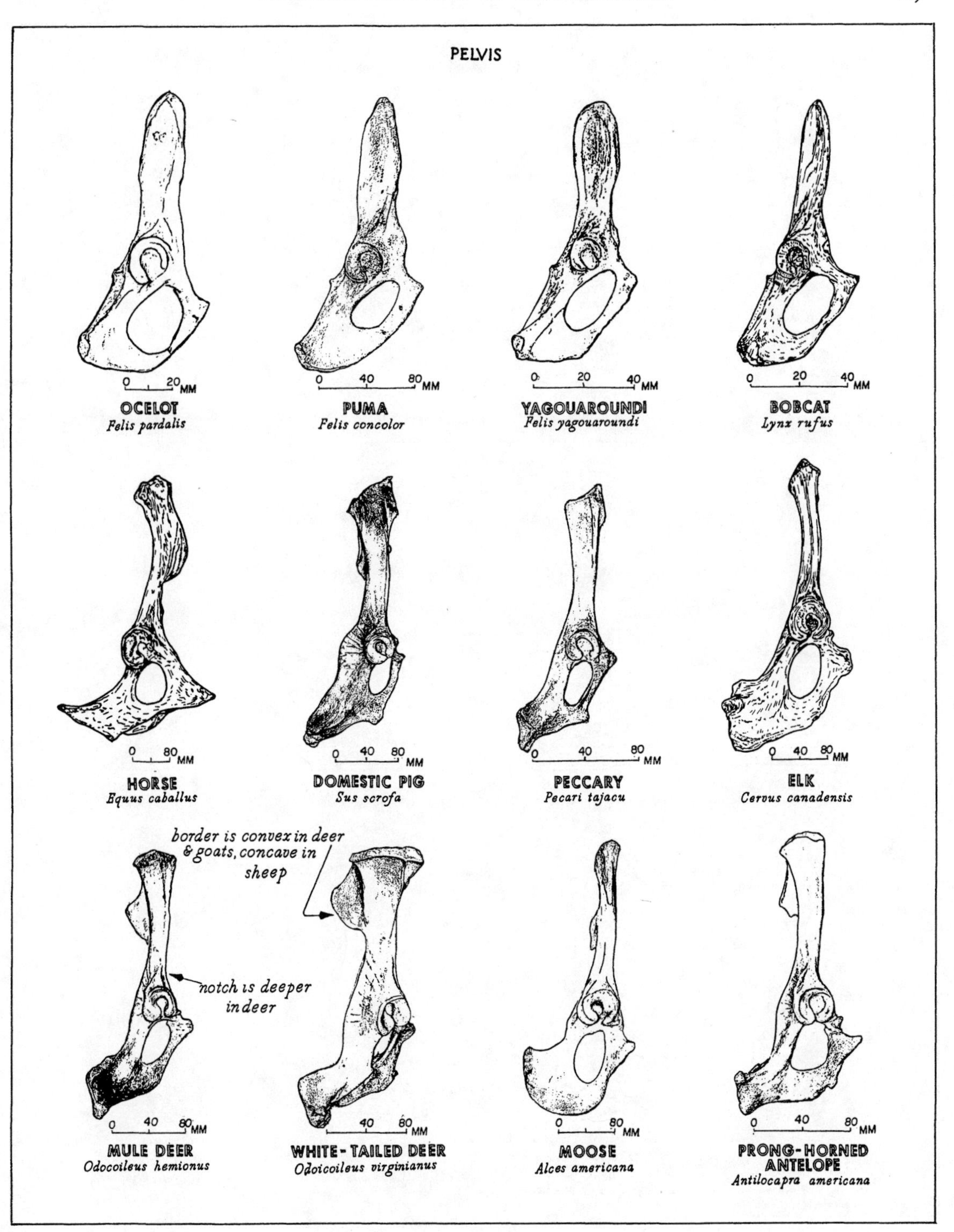
PELVIS
0 20 MM
OCELOT
Felis pardalis
0 40 80 MM
PUMA
Felis concolor
0 20 40 MM
YAGOUAROUNDI
Felis yagouaroundi
0 20 40 MM
BOBCAT
Lynx rufus
0 80 MM
HORSE
Equus caballus
0 40 80 MM
DOMESTIC PIG
Sus scrofa
0 40 80 MM
PECCARY
Pecari tajacu
0 40 80 MM
ELK
Cervus canadensis
border is convex in deer & goats, concave in sheep
notch is deeper in deer
0 40 80 MM
MULE DEER
Odocoileus hemionus
40 80 MM
WHITE-TAILED DEER
Odocoileus virginianus
0 80 MM
MOOSE
Alces americana
0 40 80 MM
PRONG-HORNED ANTELOPE
Antilocapra americana

FIGURE 76

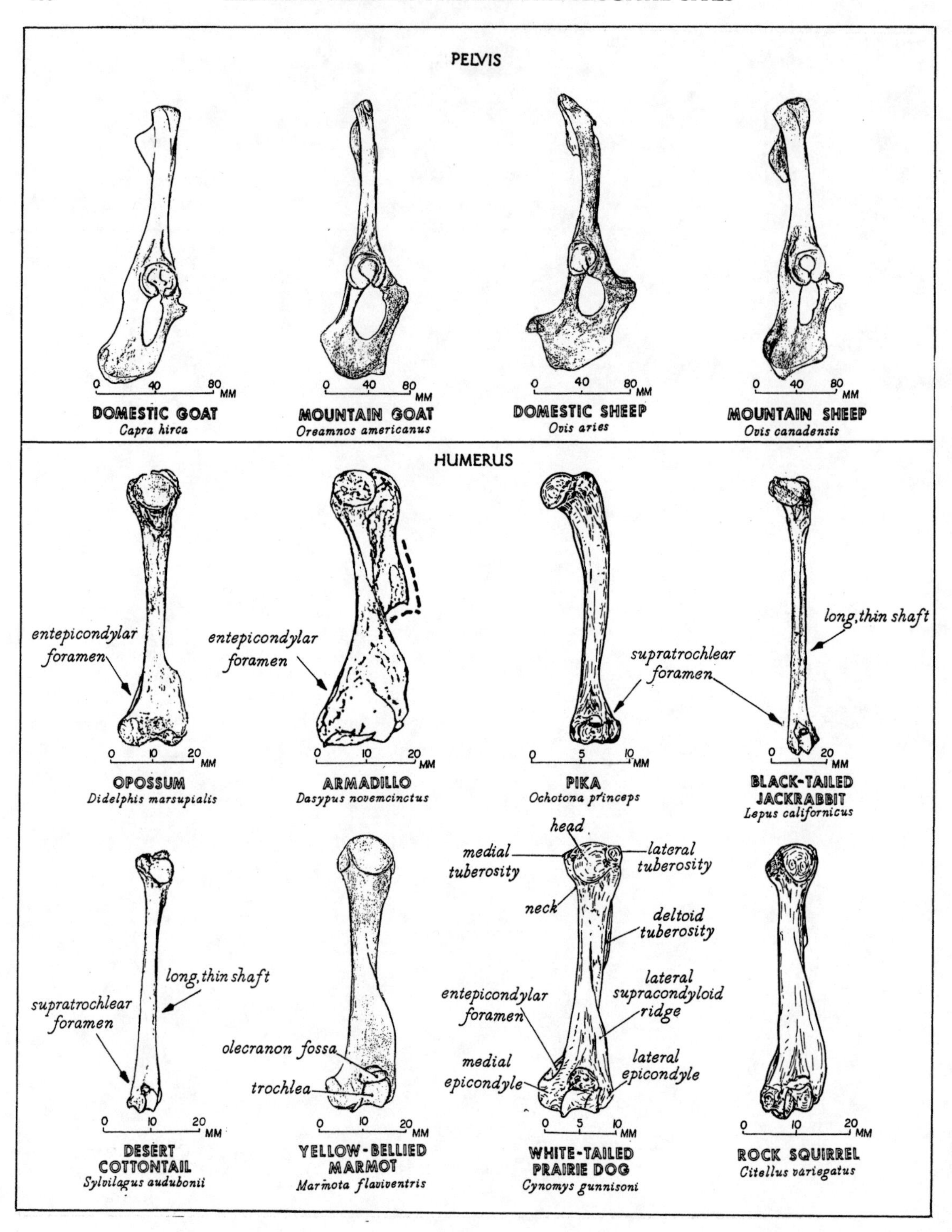
PELVIS
0 40 80 MM
DOMESTIC GOAT
Capra hirca
0 40 80 MM
MOUNTAIN GOAT
Oreamnos americanus
0 40 80 MM
DOMESTIC SHEEP
Ovis aries
0 40 80 MM
MOUNTAIN SHEEP
Ovis canadensis
HUMERUS
entepicondylar foramen
0 10 20 MM
OPOSSUM
Didelphis marsupialis
entepicondylar foramen
0 10 20 MM
ARMADILLO
Dasypus novemcinctus
supratrochlear foramen
0 5 10 MM
PIKA
Ochotona princeps
long, thin shaft
0 20 MM
BLACK-TAILED JACKRABBIT
Lepus californicus
supratrochlear foramen
long, thin shaft
0 10 20 MM
DESERT COTTONTAIL
Sylvilagus audubonii
olecranon fossa
trochlea
0 10 20 MM
YELLOW-BELLIED MARMOT
Marmota flaviventris
head
medial tuberosity
lateral tuberosity
neck
deltoid tuberosity
lateral supracondyloid ridge
entepicondylar foramen
medial epicondyle
lateral epicondyle
0 5 10 MM
WHITE-TAILED PRAIRIE DOG
Cynomys gunnisoni
0 10 20 MM
ROCK SQUIRREL
Citellus variegatus

FIGURE 77

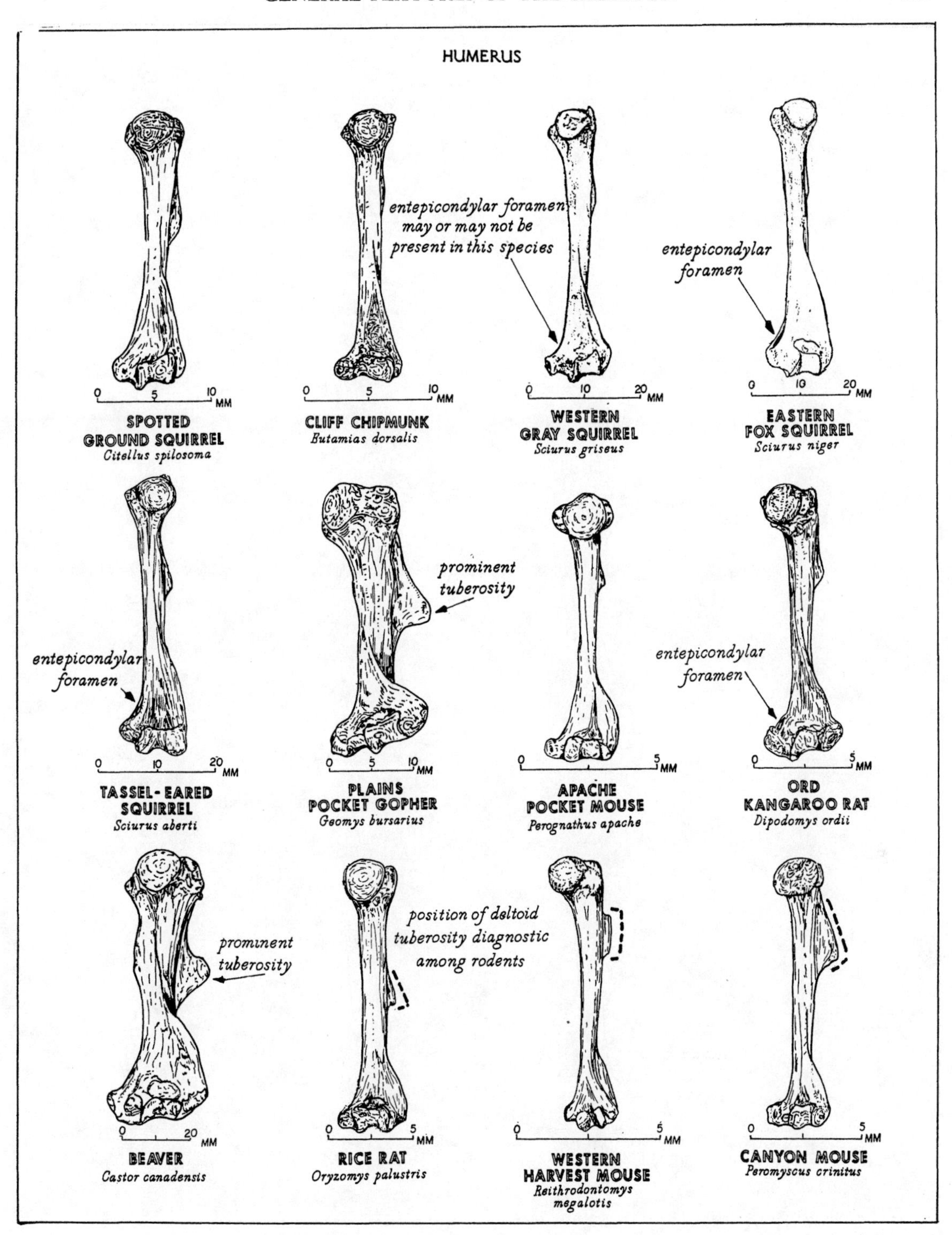
HUMERUS
entepicondylar foramen may or may not be present in this species
entepicondylar foramen
0 5 10 MM
SPOTTED GROUND SQUIRREL
Citellus spilosoma
0 5 10 MM
CLIFF CHIPMUNK
Eutamias dorsalis
0 10 20 MM
WESTERN GRAY SQUIRREL
Sciurus griseus
0 10 20 MM
EASTERN FOX SQUIRREL
Sciurus niger
prominent tuberosity
entepicondylar foramen
entepicondylar foramen
0 10 20 MM
TASSEL-EARED SQUIRREL
Sciurus aberti
0 5 10 MM
PLAINS POCKET GOPHER
Geomys bursarius
0 5 MM
APACHE POCKET MOUSE
Perognathus apache
0 5 MM
ORD KANGAROO RAT
Dipodomys ordii
prominent tuberosity
position of deltoid tuberosity diagnostic among rodents
0 20 MM
BEAVER
Castor canadensis
0 5 MM
RICE RAT
Oryzomys palustris
0 5 MM
WESTERN HARVEST MOUSE
Reithrodontomys megalotis
0 5 MM
CANYON MOUSE
Peromyscus crinitus

FIGURE 78

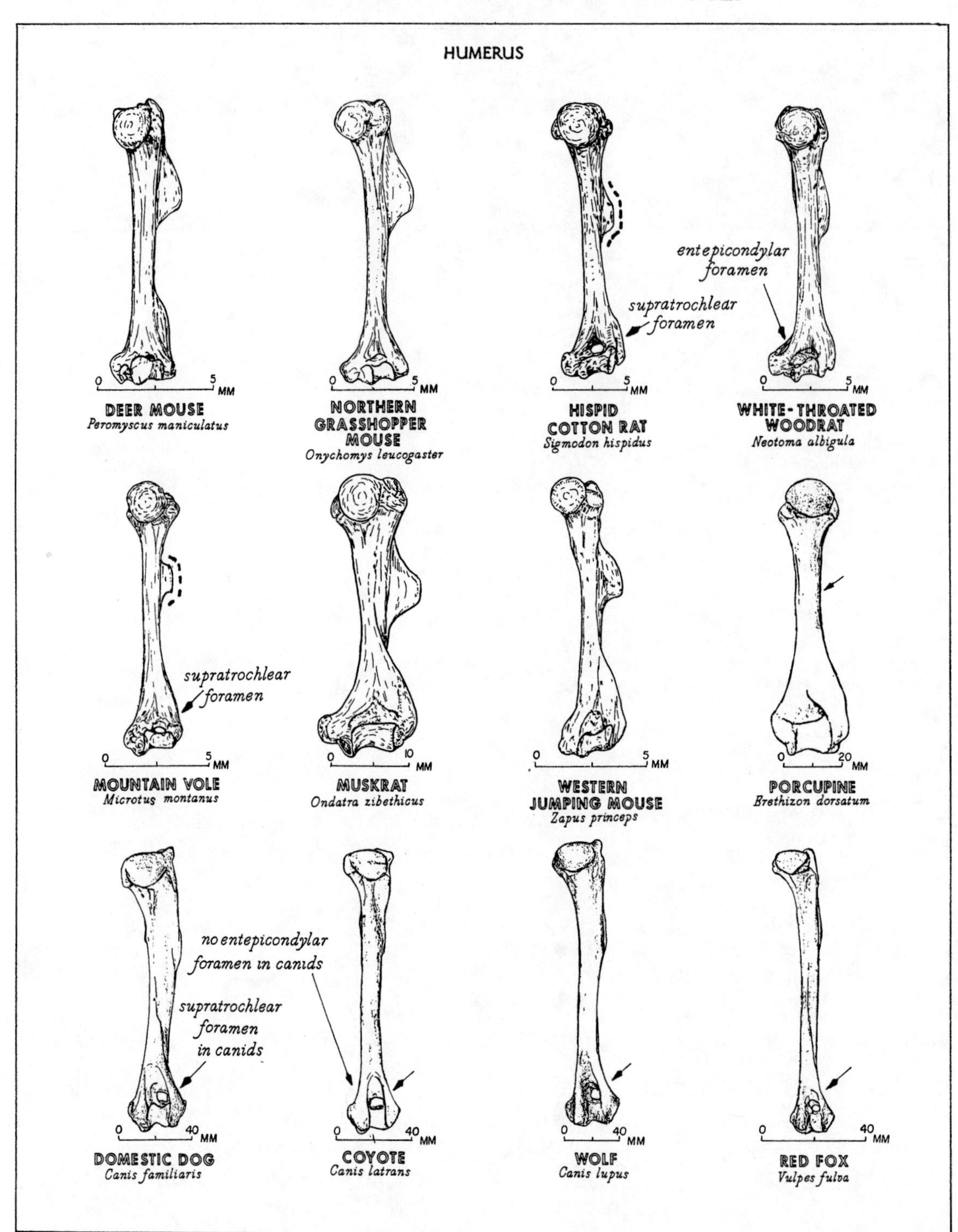
HUMERUS
0 5 MM
DEER MOUSE
Peromyscus maniculatus
0 5 MM
NORTHERN GRASSHOPPER MOUSE
Onychomys leucogaster
supratrochlear foramen
0 5 MM
HISPID COTTON RAT
Sigmodon hispidus
entepicondylar foramen
0 5 MM
WHITE-THROATED WOODRAT
Neotoma albigula
supratrochlear foramen
0 5 MM
MOUNTAIN VOLE
Microtus montanus
0 10 MM
MUSKRAT
Ondatra zibethicus
0 5 MM
WESTERN JUMPING MOUSE
Zapus princeps
0 20 MM
PORCUPINE
Erethizon dorsatum
no entepicondylar foramen in canids
supratrochlear foramen in canids
0 40 MM
DOMESTIC DOG
Canis familiaris
0 40 MM
COYOTE
Canis latrans
0 40 MM
WOLF
Canis lupus
0 40 MM
RED FOX
Vulpes fulva

FIGURE 79

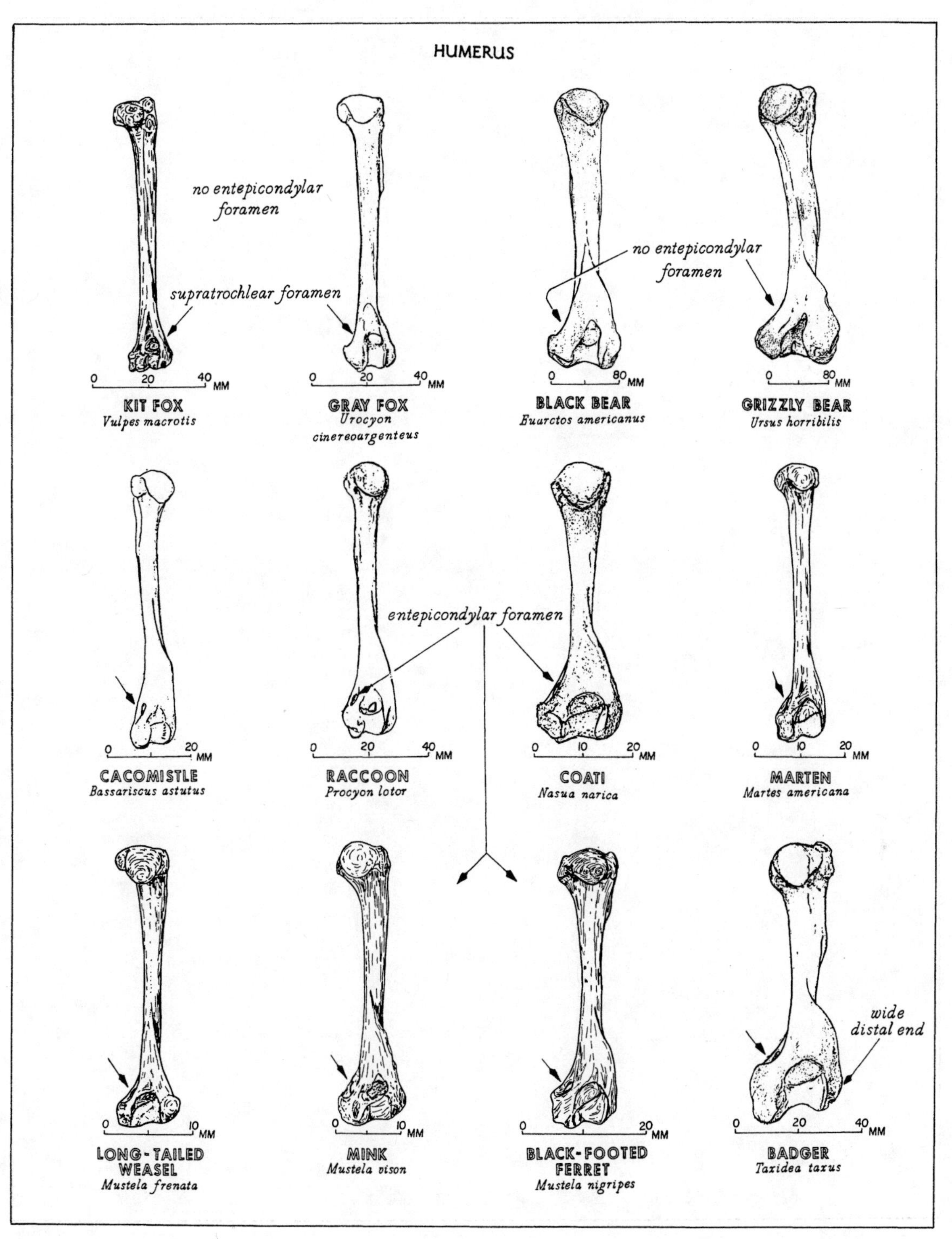
HUMERUS
no entepicondylar foramen
supratrochlear foramen
0 20 40 MM
KIT FOX
Vulpes macrotis
0 20 40 MM
GRAY FOX
Urocyon cinereoargenteus
no entepicondylar foramen
0 80 MM
BLACK BEAR
Euarctos americanus
0 80 MM
GRIZZLY BEAR
Ursus horribilis
entepicondylar foramen
0 20 MM
CACOMISTLE
Bassariscus astutus
0 20 40 MM
RACCOON
Procyon lotor
0 10 20 MM
COATI
Nasua narica
0 10 20 MM
MARTEN
Martes americana
wide distal end
0 10 MM
LONG-TAILED WEASEL
Mustela frenata
0 10 MM
MINK
Mustela vison
0 20 MM
BLACK-FOOTED FERRET
Mustela nigripes
0 20 40 MM
BADGER
Taxidea taxus

FIGURE 80

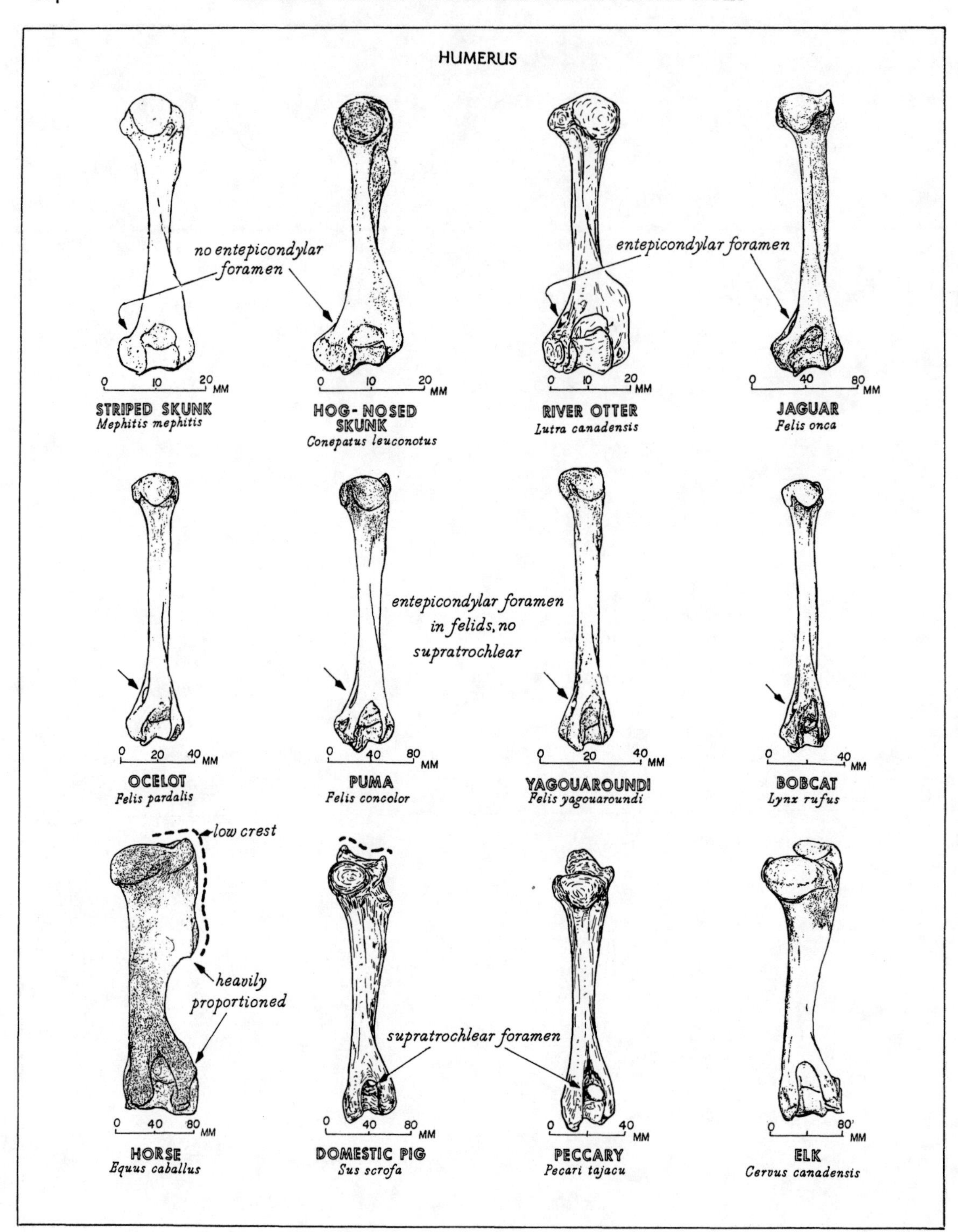
HUMERUS
no entepicondylar foramen
entepicondylar foramen
0 10 20 MM
STRIPED SKUNK
Mephitis mephitis
0 10 20 MM
HOG-NOSED SKUNK
Conepatus leuconotus
0 10 20 MM
RIVER OTTER
Lutra canadensis
0 40 80 MM
JAGUAR
Felis onca
entepicondylar foramen in felids, no supratrochlear
0 20 40 MM
OCELOT
Felis pardalis
0 40 80 MM
PUMA
Felis concolor
0 20 40 MM
YAGOUAROUNDI
Felis yagouaroundi
0 40 MM
BOBCAT
Lynx rufus
low crest
heavily proportioned
supratrochlear foramen
0 40 80 MM
HORSE
Equus caballus
0 40 80 MM
DOMESTIC PIG
Sus scrofa
0 40 MM
PECCARY
Pecari tajacu
0 80 MM
ELK
Cervus canadensis

FIGURE 81

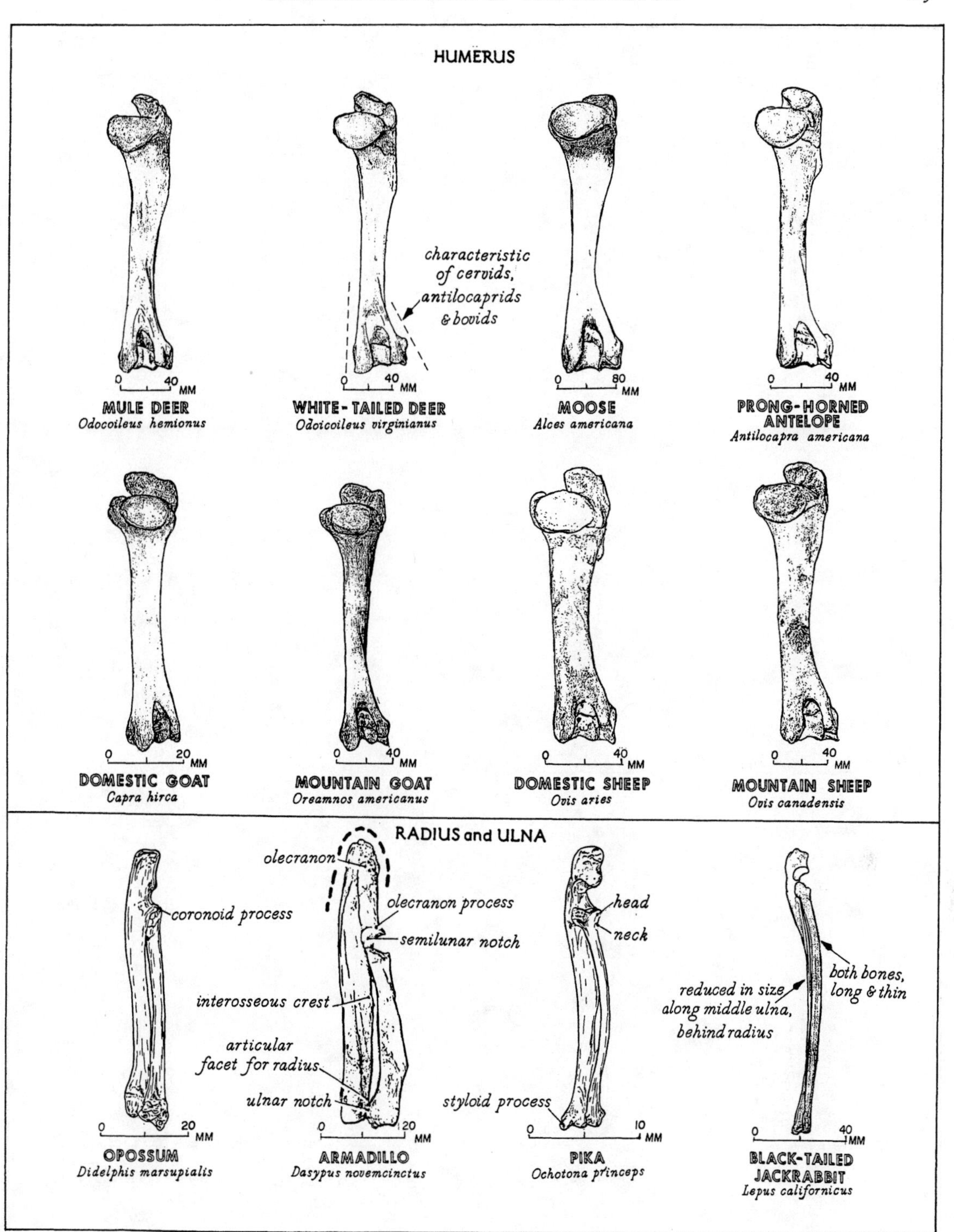
HUMERUS
characteristic of cervids, antilocaprids & bovids
0 40 MM
MULE DEER
Odocoileus hemionus
0 40 MM
WHITE-TAILED DEER
Odocoileus virginianus
0 80 MM
MOOSE
Alces americana
0 40 MM
PRONG-HORNED ANTELOPE
Antilocapra americana
0 20 MM
DOMESTIC GOAT
Capra hirca
0 40 MM
MOUNTAIN GOAT
Oreamnos americanus
0 40 MM
DOMESTIC SHEEP
Ovis aries
0 40 MM
MOUNTAIN SHEEP
Ovis canadensis
RADIUS and ULNA
coronoid process
olecranon
olecranon process
semilunar notch
interosseous crest
articular facet for radius
ulnar notch
head
neck
styloid process
reduced in size along middle ulna, behind radius
both bones, long & thin
0 20 MM
OPOSSUM
Didelphis marsupialis
0 20 MM
ARMADILLO
Dasypus novemcinctus
0 10 MM
PIKA
Ochotona princeps
0 40 MM
BLACK-TAILED JACKRABBIT
Lepus californicus

FIGURE 82

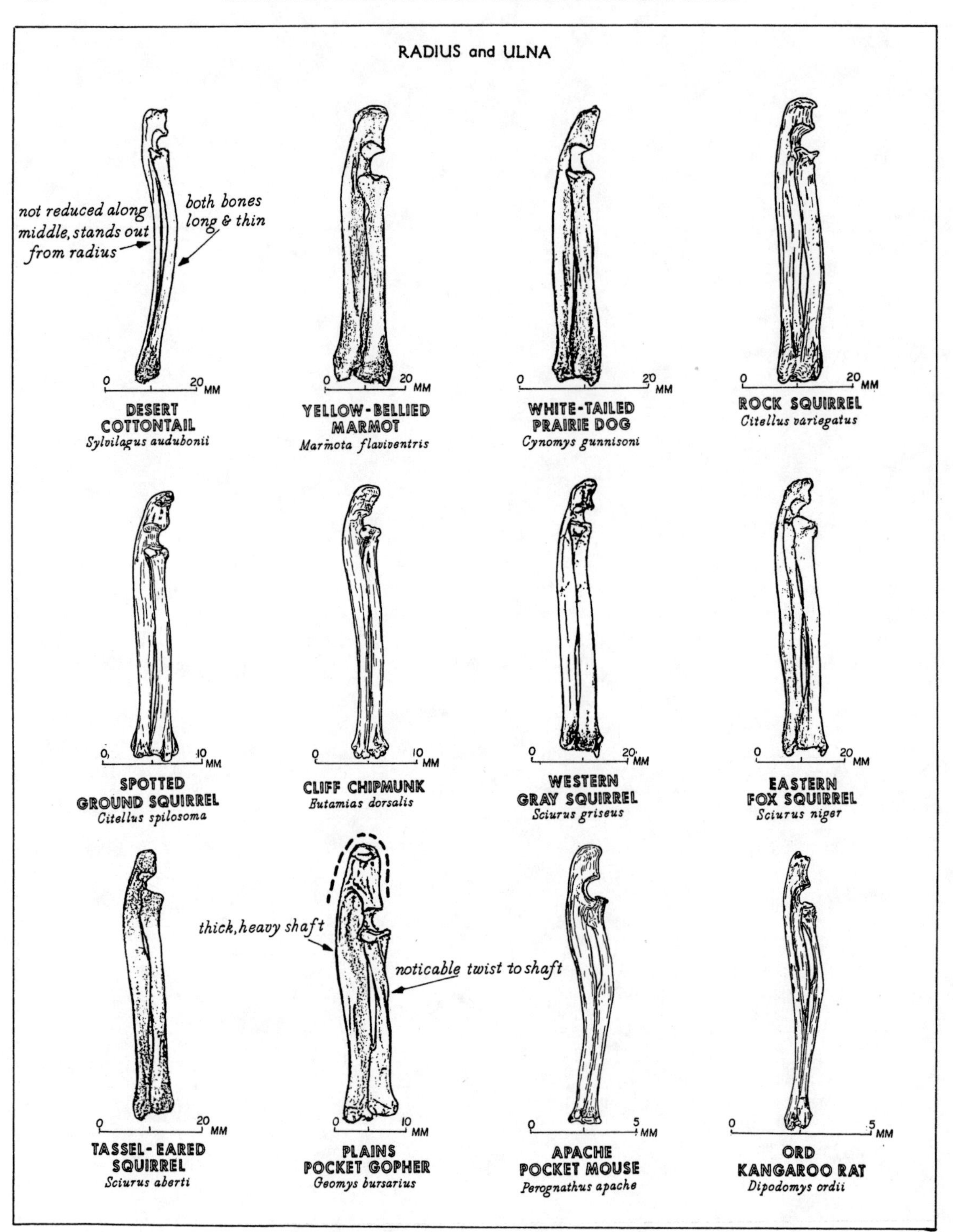
RADIUS and ULNA
not reduced along middle, stands out from radius
both bones long & thin
0 20 MM
DESERT COTTONTAIL
Sylvilagus audubonii
0 20 MM
YELLOW-BELLIED MARMOT
Marmota flaviventris
0 20 MM
WHITE-TAILED PRAIRIE DOG
Cynomys gunnisoni
0 20 MM
ROCK SQUIRREL
Citellus variegatus
0 10 MM
SPOTTED GROUND SQUIRREL
Citellus spilosoma
0 10 MM
CLIFF CHIPMUNK
Eutamias dorsalis
0 20 MM
WESTERN GRAY SQUIRREL
Sciurus griseus
0 20 MM
EASTERN FOX SQUIRREL
Sciurus niger
0 20 MM
TASSEL-EARED SQUIRREL
Sciurus aberti
thick, heavy shaft
noticable twist to shaft
0 10 MM
PLAINS POCKET GOPHER
Geomys bursarius
0 5 MM
APACHE POCKET MOUSE
Perognathus apache
0 5 MM
ORD KANGAROO RAT
Dipodomys ordii

FIGURE 83

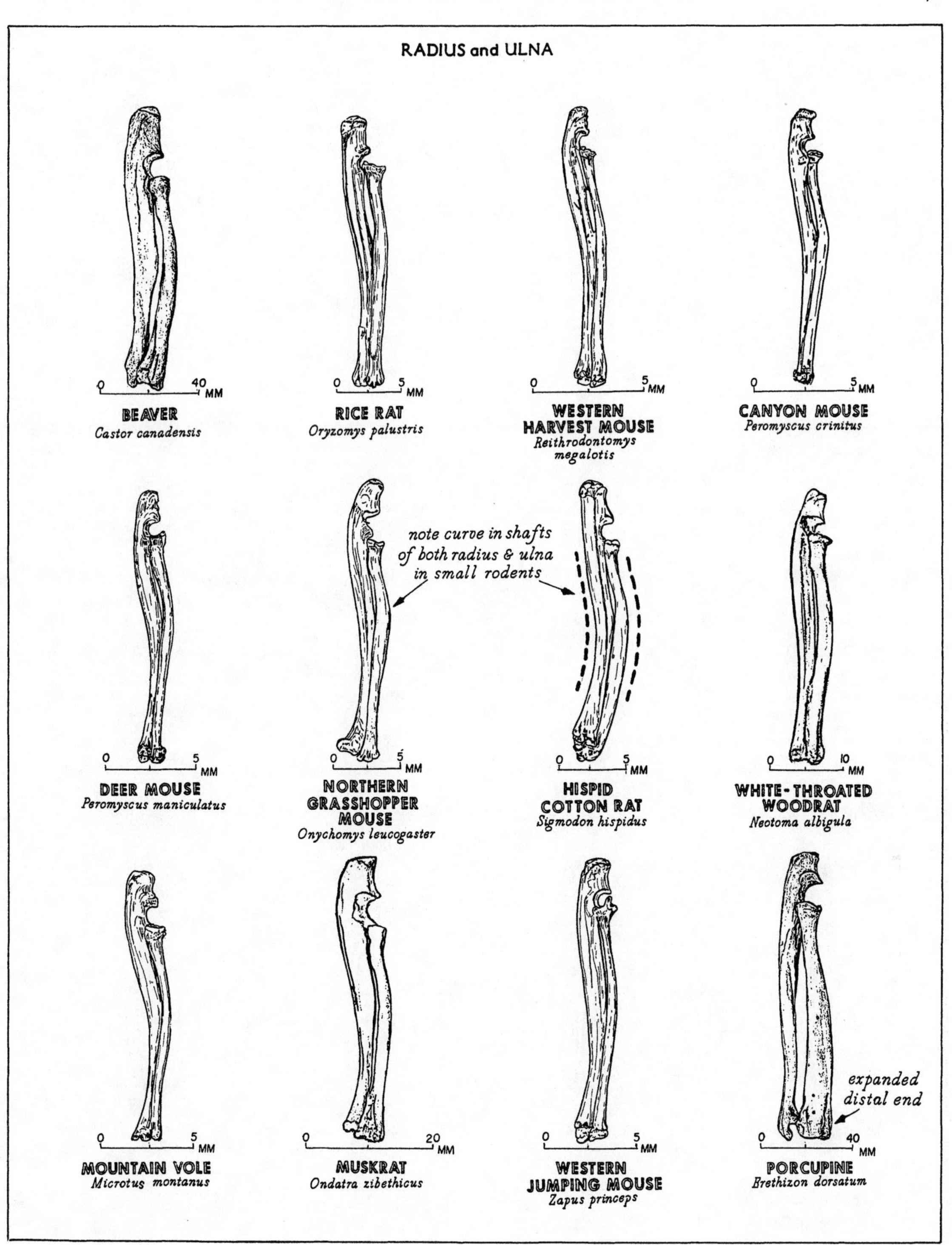
RADIUS and ULNA
0 40 MM
BEAVER
Castor canadensis
0 5 MM
RICE RAT
Oryzomys palustris
0 5 MM
WESTERN HARVEST MOUSE
Reithrodontomys megalotis
0 5 MM
CANYON MOUSE
Peromyscus crinitus
0 5 MM
DEER MOUSE
Peromyscus maniculatus
0 5 MM
NORTHERN GRASSHOPPER MOUSE
Onychomys leucogaster
note curve in shafts of both radius & ulna in small rodents
0 5 MM
HISPID COTTON RAT
Sigmodon hispidus
0 10 MM
WHITE-THROATED WOODRAT
Neotoma albigula
0 5 MM
MOUNTAIN VOLE
Microtus montanus
0 20 MM
MUSKRAT
Ondatra zibethicus
0 5 MM
WESTERN JUMPING MOUSE
Zapus princeps
expanded distal end
0 40 MM
PORCUPINE
Erethizon dorsatum

FIGURE 84

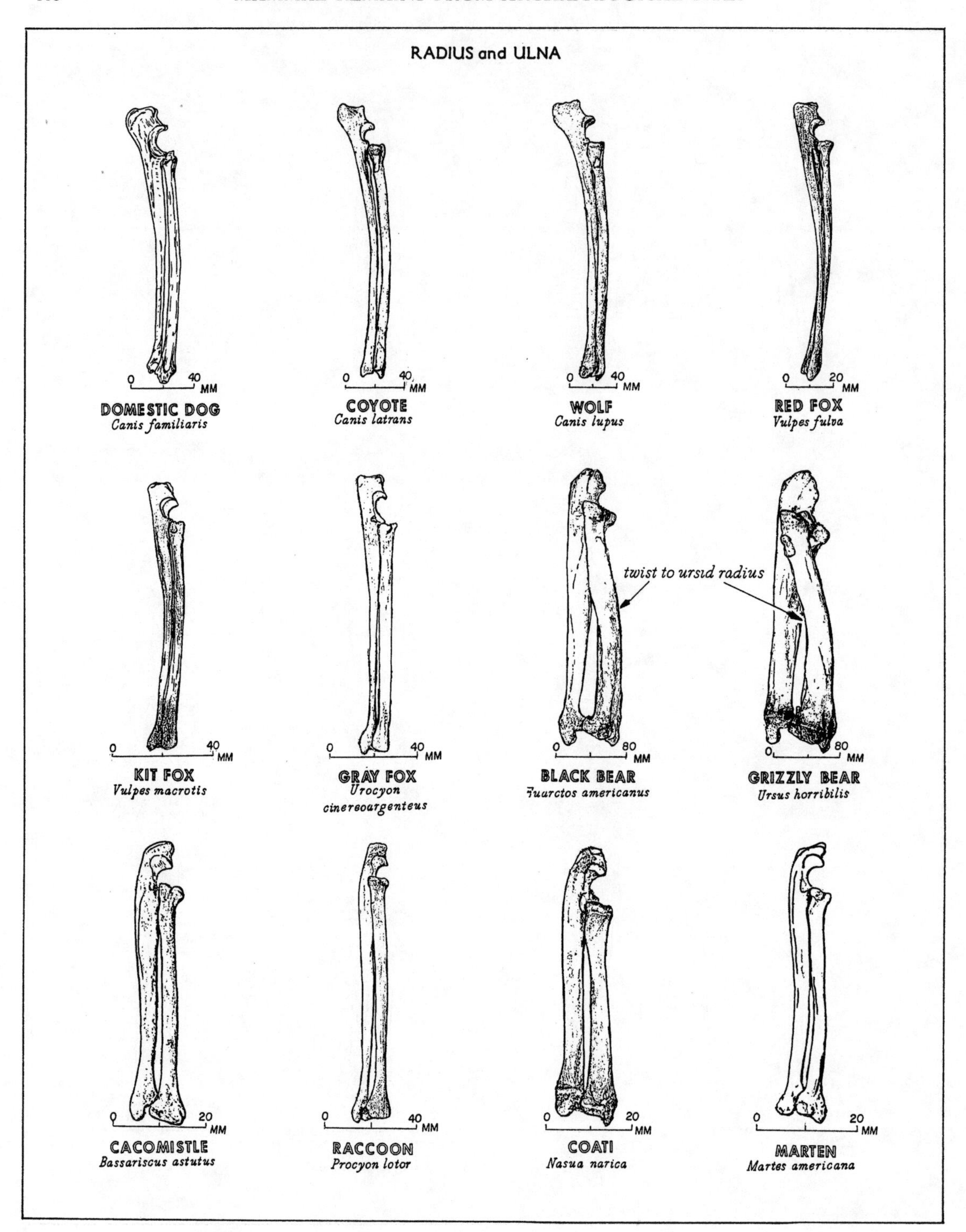
RADIUS and ULNA
0 40 MM
DOMESTIC DOG
Canis familiaris
0 40 MM
COYOTE
Canis latrans
0 40 MM
WOLF
Canis lupus
0 20 MM
RED FOX
Vulpes fulva
0 40 MM
KIT FOX
Vulpes macrotis
0 40 MM
GRAY FOX
Urocyon
cinereoargenteus
twist to ursid radius
0 80 MM
BLACK BEAR
Euarctos americanus
0 80 MM
GRIZZLY BEAR
Ursus horribilis
0 20 MM
CACOMISTLE
Bassariscus astutus
0 40 MM
RACCOON
Procyon lotor
0 20 MM
COATI
Nasua narica
0 20 MM
MARTEN
Martes americana

FIGURE 85

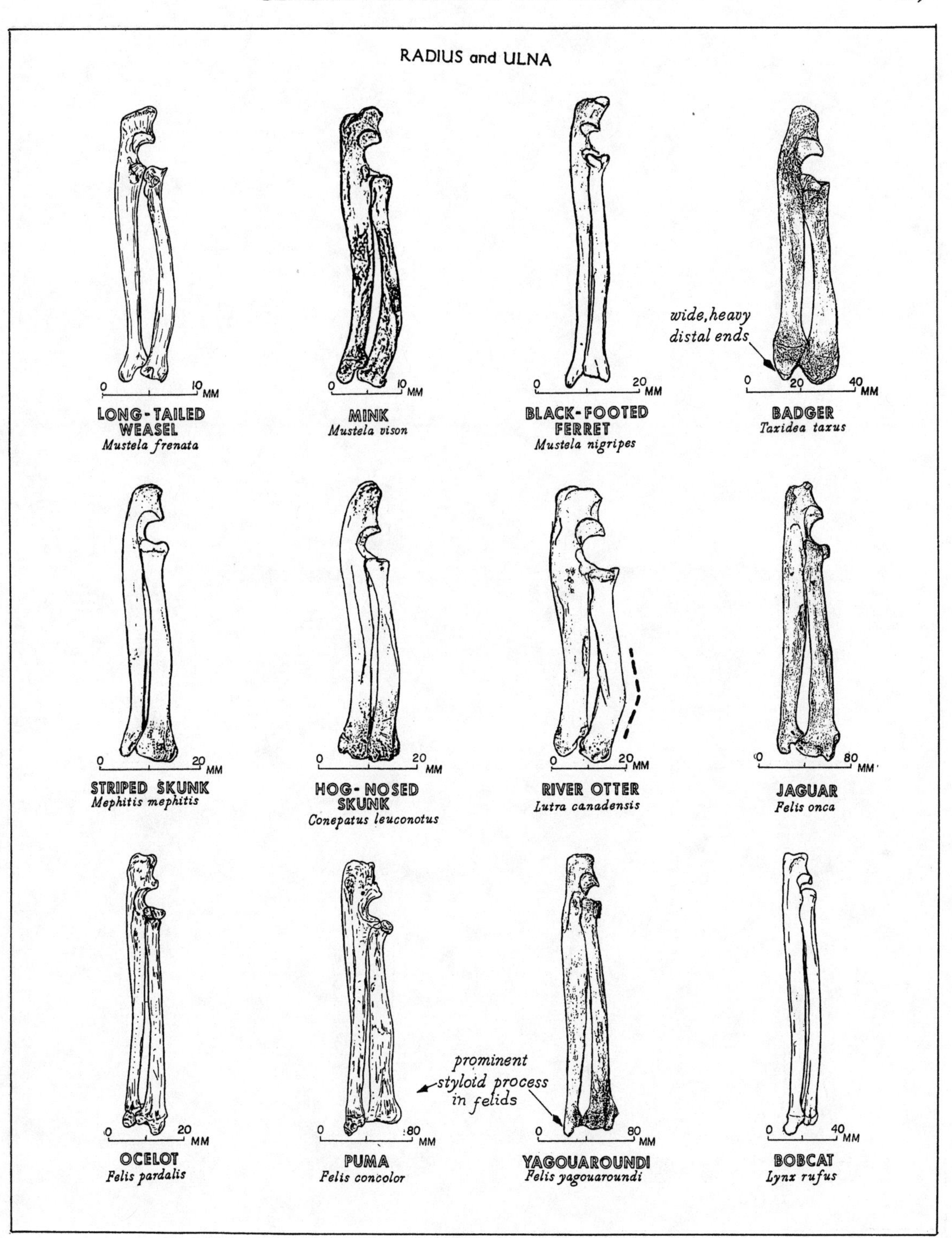
RADIUS and ULNA
0 10 MM
LONG-TAILED WEASEL
Mustela frenata
0 10 MM
MINK
Mustela vison
0 20 MM
BLACK-FOOTED FERRET
Mustela nigripes
wide, heavy distal ends
0 20 40 MM
BADGER
Taxidea taxus
0 20 MM
STRIPED SKUNK
Mephitis mephitis
0 20 MM
HOG-NOSED SKUNK
Conepatus leuconotus
0 20 MM
RIVER OTTER
Lutra canadensis
0 80 MM
JAGUAR
Felis onca
0 20 MM
OCELOT
Felis pardalis
0 80 MM
PUMA
Felis concolor
prominent styloid process in felids
0 80 MM
YAGOUAROUNDI
Felis yagouaroundi
0 40 MM
BOBCAT
Lynx rufus

FIGURE 86

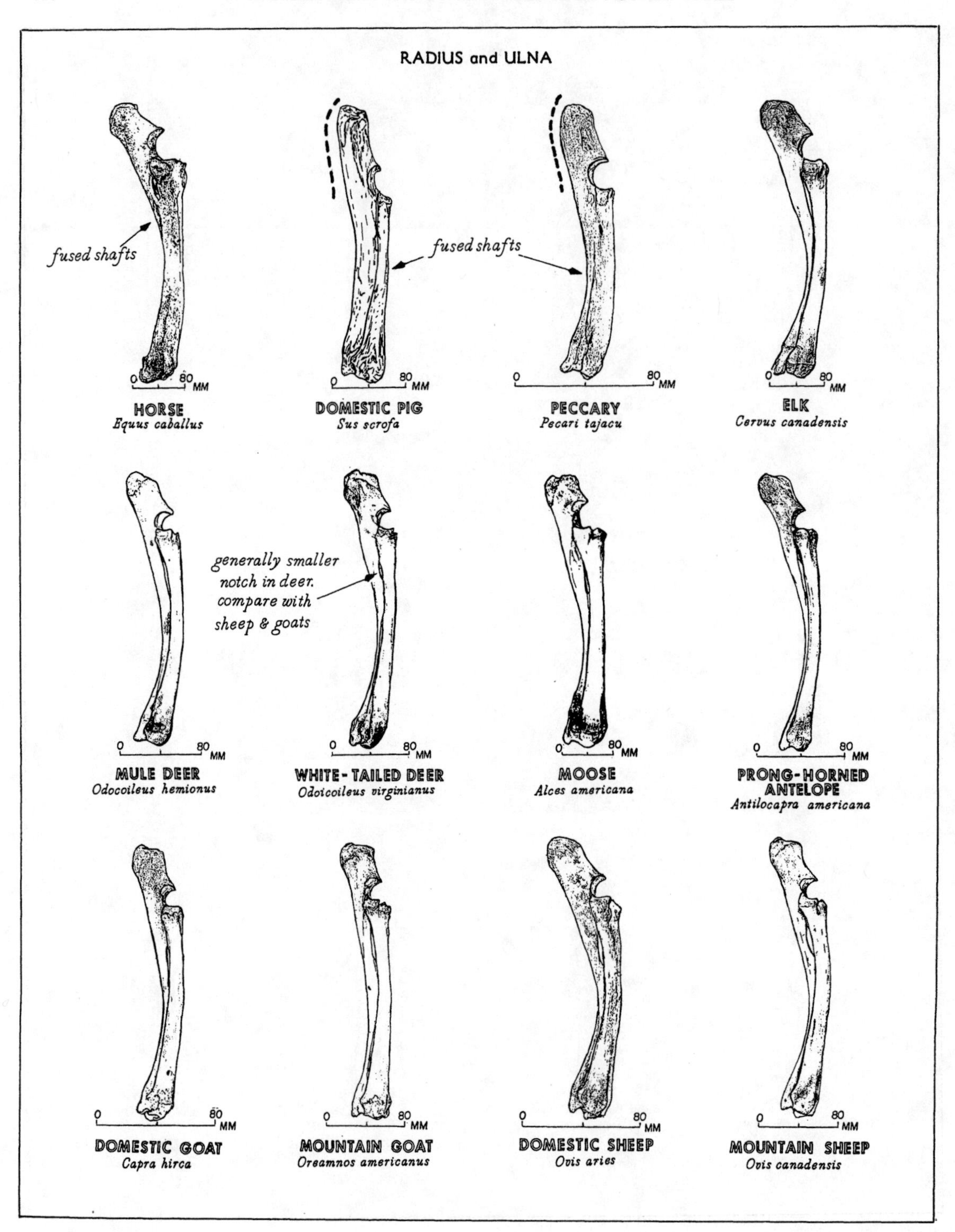
RADIUS and ULNA
fused shafts
fused shafts
0 80 MM
HORSE
Equus caballus
DOMESTIC PIG
Sus scrofa
PECCARY
Pecari tajacu
ELK
Cervus canadensis
generally smaller notch in deer. compare with sheep & goats
MULE DEER
Odocoileus hemionus
WHITE-TAILED DEER
Odocoileus virginianus
MOOSE
Alces americana
PRONG-HORNED ANTELOPE
Antilocapra americana
DOMESTIC GOAT
Capra hirca
MOUNTAIN GOAT
Oreamnos americanus
DOMESTIC SHEEP
Ovis aries
MOUNTAIN SHEEP
Ovis canadensis

FIGURE 87

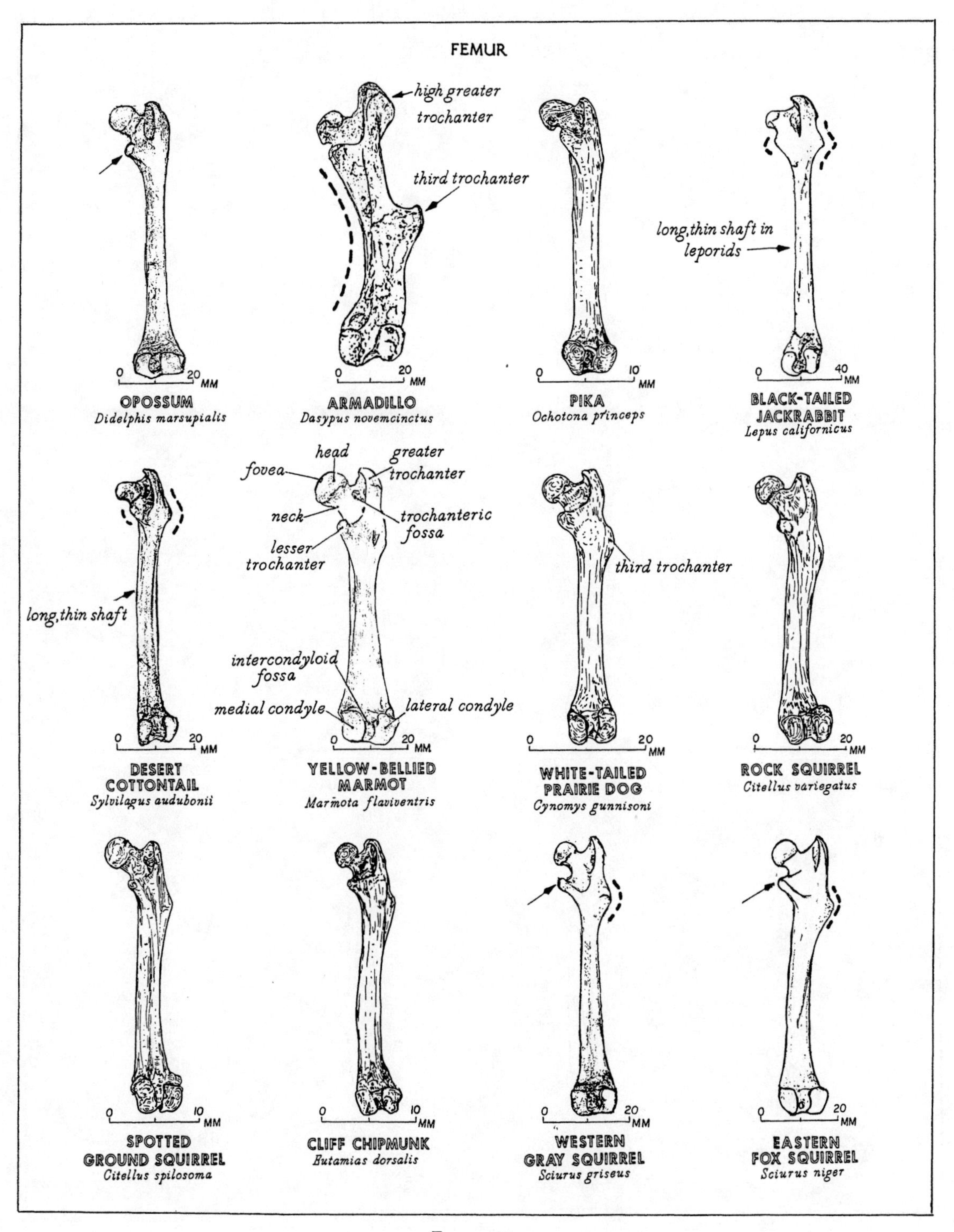
FEMUR
high greater
trochanter
third trochanter
long, thin shaft in
leporids
0 20 MM
OPOSSUM
Didelphis marsupialis
0 20 MM
ARMADILLO
Dasypus novemcinctus
0 10 MM
PIKA
Ochotona princeps
0 40 MM
BLACK-TAILED
JACKRABBIT
Lepus californicus
head
greater
trochanter
fovea
neck
trochanteric
fossa
lesser
trochanter
third trochanter
long, thin shaft
intercondyloid
fossa
medial condyle
lateral condyle
0 20 MM
DESERT
COTTONTAIL
Sylvilagus audubonii
0 20 MM
YELLOW-BELLIED
MARMOT
Marmota flaviventris
0 20 MM
WHITE-TAILED
PRAIRIE DOG
Cynomys gunnisoni
0 20 MM
ROCK SQUIRREL
Citellus variegatus
0 10 MM
SPOTTED
GROUND SQUIRREL
Citellus spilosoma
0 10 MM
CLIFF CHIPMUNK
Eutamias dorsalis
0 20 MM
WESTERN
GRAY SQUIRREL
Sciurus griseus
0 20 MM
EASTERN
FOX SQUIRREL
Sciurus niger

FIGURE 88

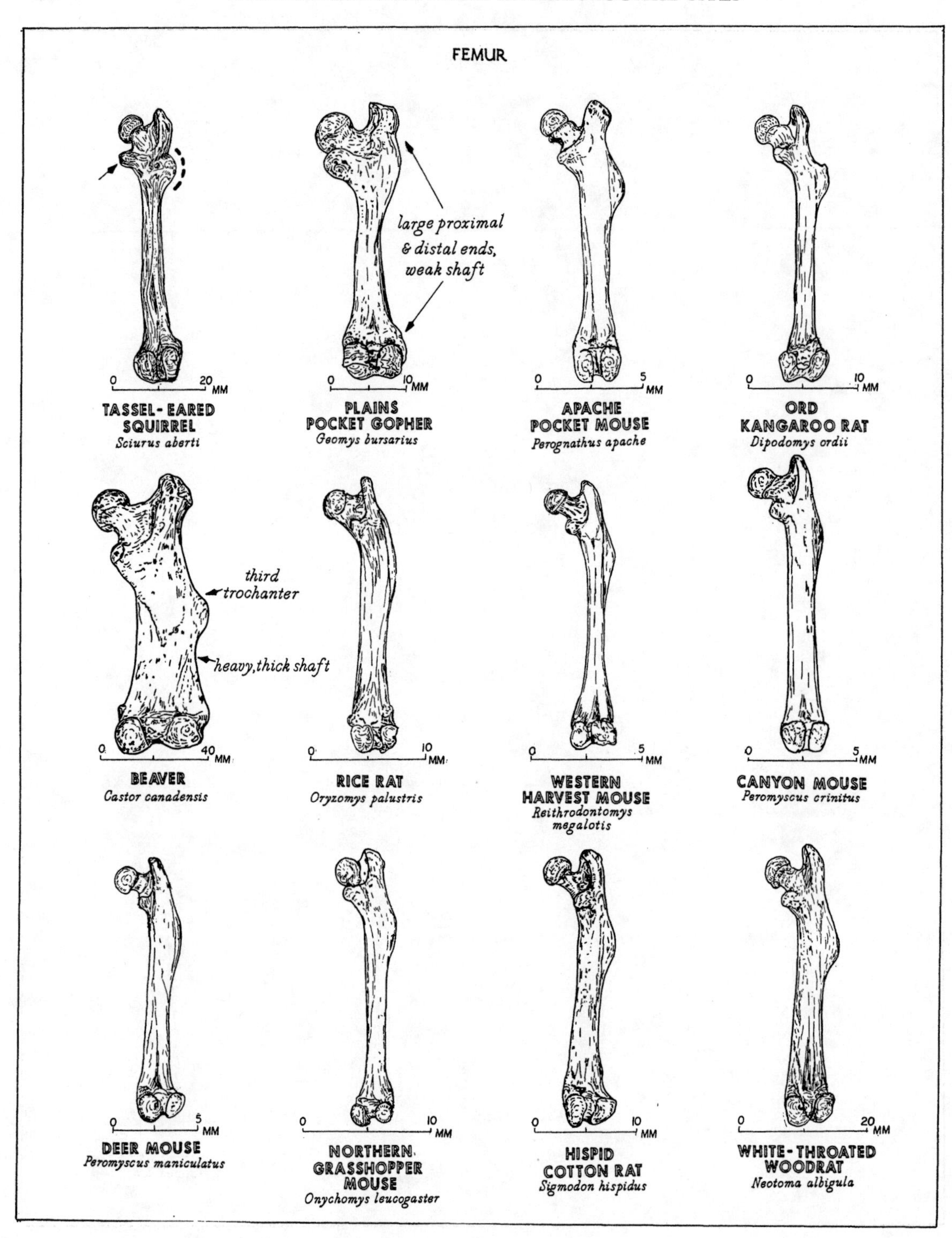
FEMUR
large proximal & distal ends, weak shaft
0 20 MM
TASSEL-EARED SQUIRREL
Sciurus aberti
0 10 MM
PLAINS POCKET GOPHER
Geomys bursarius
0 5 MM
APACHE POCKET MOUSE
Perognathus apache
0 10 MM
ORD KANGAROO RAT
Dipodomys ordii
third trochanter
heavy, thick shaft
0 40 MM
BEAVER
Castor canadensis
0 10 MM
RICE RAT
Oryzomys palustris
0 5 MM
WESTERN HARVEST MOUSE
Reithrodontomys megalotis
0 5 MM
CANYON MOUSE
Peromyscus crinitus
0 5 MM
DEER MOUSE
Peromyscus maniculatus
0 10 MM
NORTHERN GRASSHOPPER MOUSE
Onychomys leucogaster
0 10 MM
HISPID COTTON RAT
Sigmodon hispidus
0 20 MM
WHITE-THROATED WOODRAT
Neotoma albigula

FIGURE 89

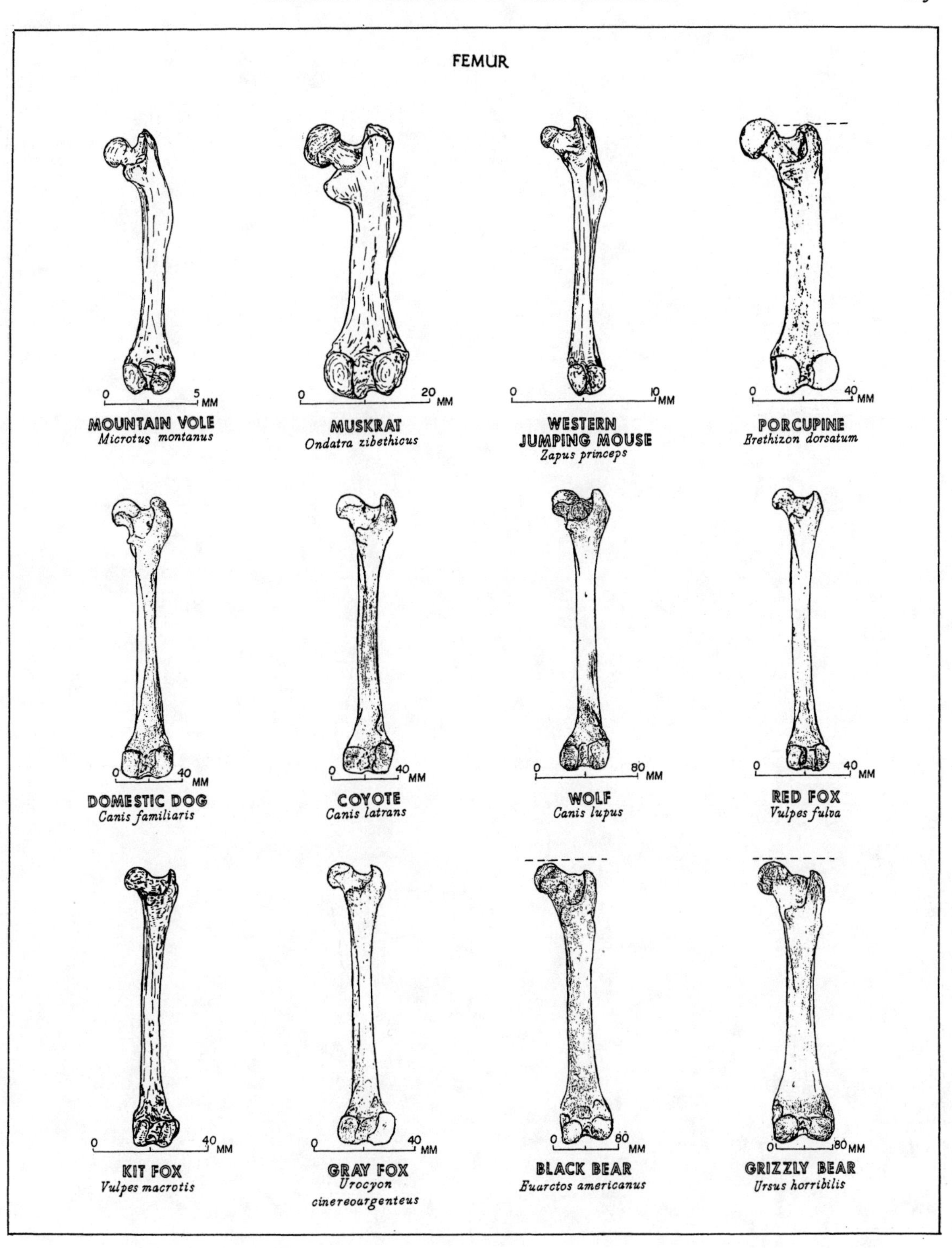
FEMUR
0 5 MM
MOUNTAIN VOLE
Microtus montanus
0 20 MM
MUSKRAT
Ondatra zibethicus
0 10 MM
WESTERN JUMPING MOUSE
Zapus princeps
0 40 MM
PORCUPINE
Erethizon dorsatum
0 40 MM
DOMESTIC DOG
Canis familiaris
0 40 MM
COYOTE
Canis latrans
0 80 MM
WOLF
Canis lupus
0 40 MM
RED FOX
Vulpes fulva
0 40 MM
KIT FOX
Vulpes macrotis
0 40 MM
GRAY FOX
Urocyon cinereoargenteus
0 80 MM
BLACK BEAR
Euarctos americanus
0 80 MM
GRIZZLY BEAR
Ursus horribilis

FIGURE 90

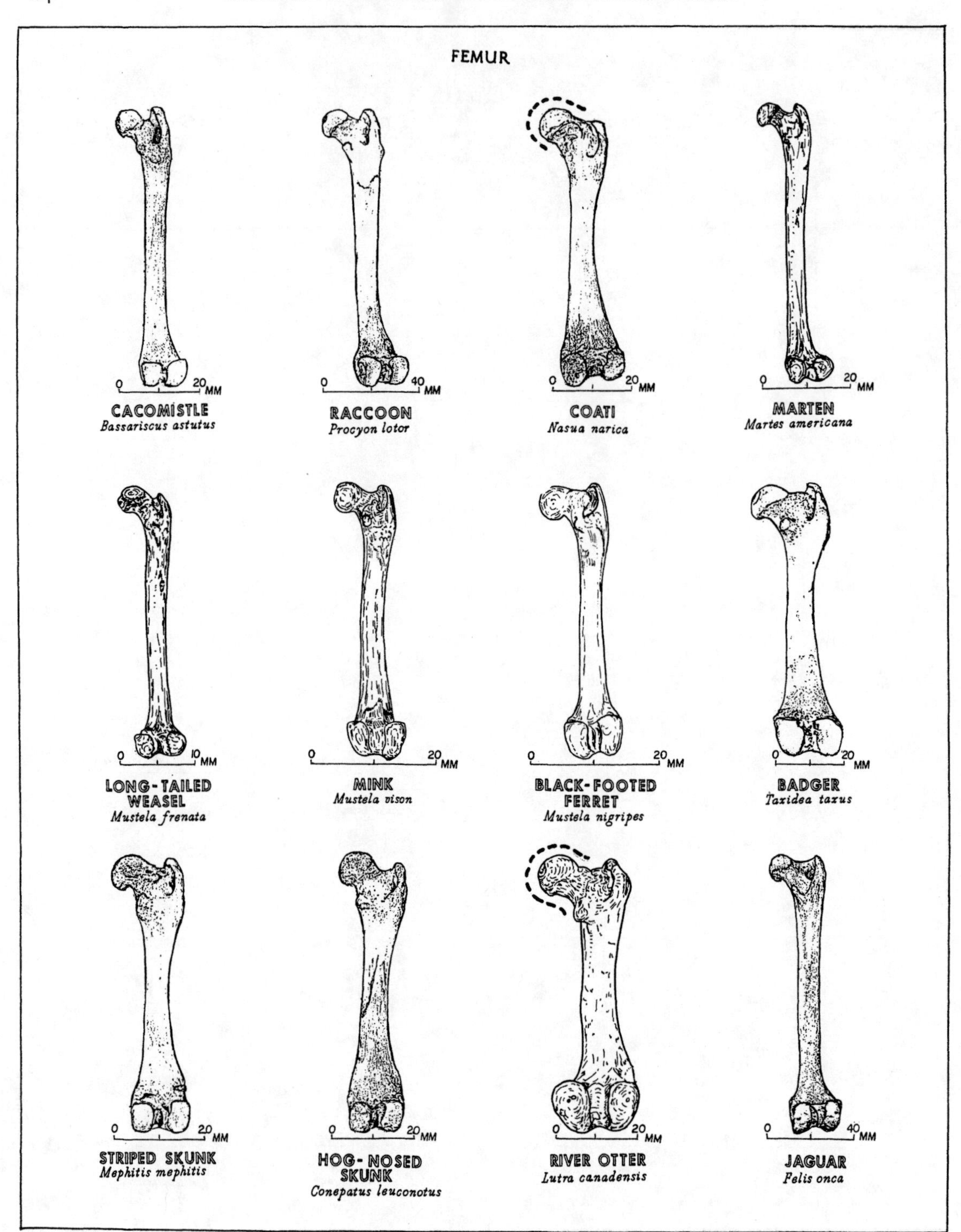
FEMUR
0 20 MM
CACOMISTLE
Bassariscus astutus
0 40 MM
RACCOON
Procyon lotor
0 20 MM
COATI
Nasua narica
0 20 MM
MARTEN
Martes americana
0 10 MM
LONG-TAILED WEASEL
Mustela frenata
0 20 MM
MINK
Mustela vison
0 20 MM
BLACK-FOOTED FERRET
Mustela nigripes
0 20 MM
BADGER
Taxidea taxus
0 20 MM
STRIPED SKUNK
Mephitis mephitis
0 20 MM
HOG-NOSED SKUNK
Conepatus leuconotus
0 20 MM
RIVER OTTER
Lutra canadensis
0 40 MM
JAGUAR
Felis onca

FIGURE 91

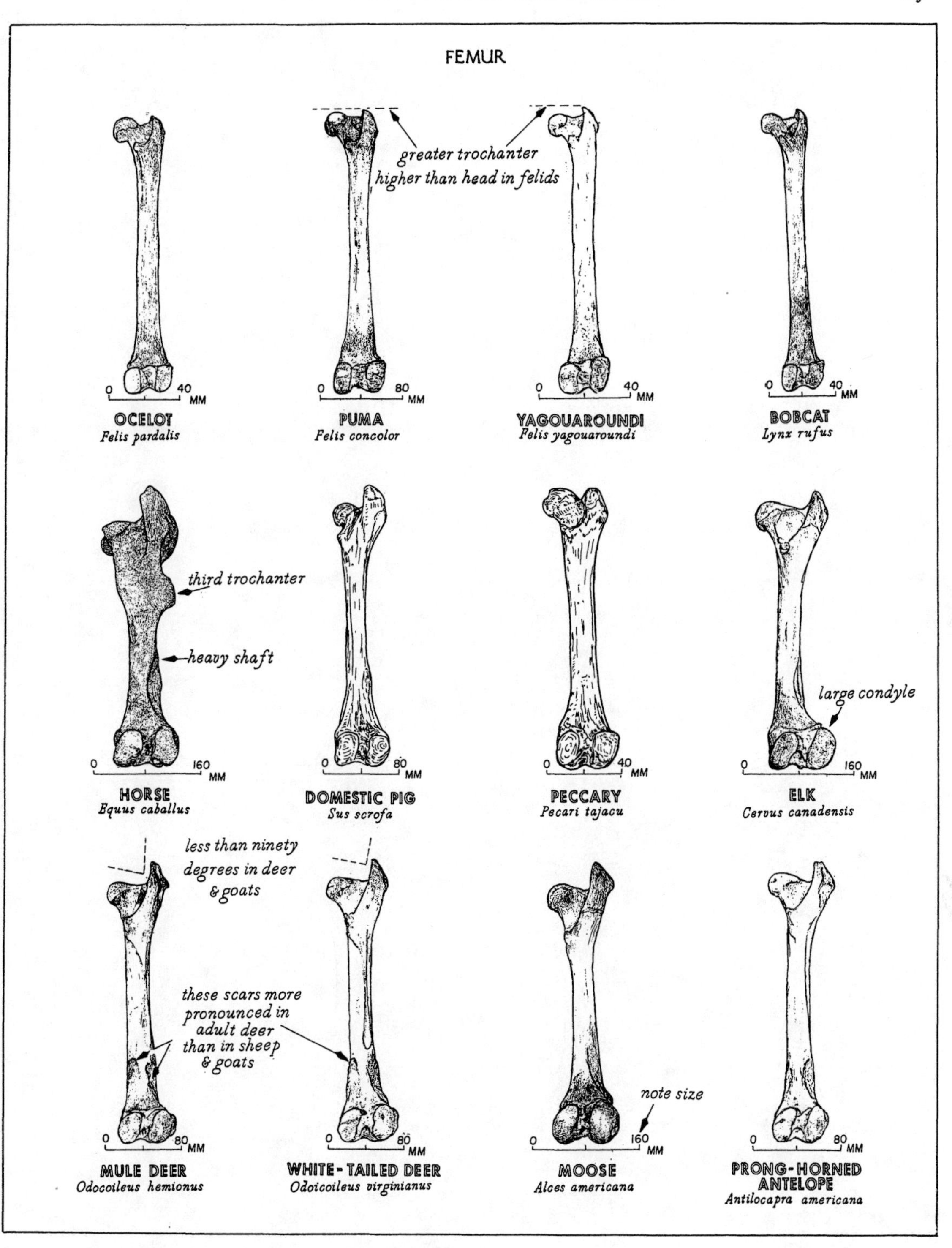
FEMUR
greater trochanter higher than head in felids
0 40 MM
OCELOT
Felis pardalis
0 80 MM
PUMA
Felis concolor
0 40 MM
YAGOUAROUNDI
Felis yagouaroundi
0 40 MM
BOBCAT
Lynx rufus
third trochanter
heavy shaft
0 160 MM
HORSE
Equus caballus
0 80 MM
DOMESTIC PIG
Sus scrofa
0 40 MM
PECCARY
Pecari tajacu
large condyle
0 160 MM
ELK
Cervus canadensis
less than ninety degrees in deer & goats
these scars more pronounced in adult deer than in sheep & goats
0 80 MM
MULE DEER
Odocoileus hemionus
0 80 MM
WHITE-TAILED DEER
Odoicoileus virginianus
note size
0 160 MM
MOOSE
Alces americana
0 80 MM
PRONG-HORNED ANTELOPE
Antilocapra americana

FIGURE 92

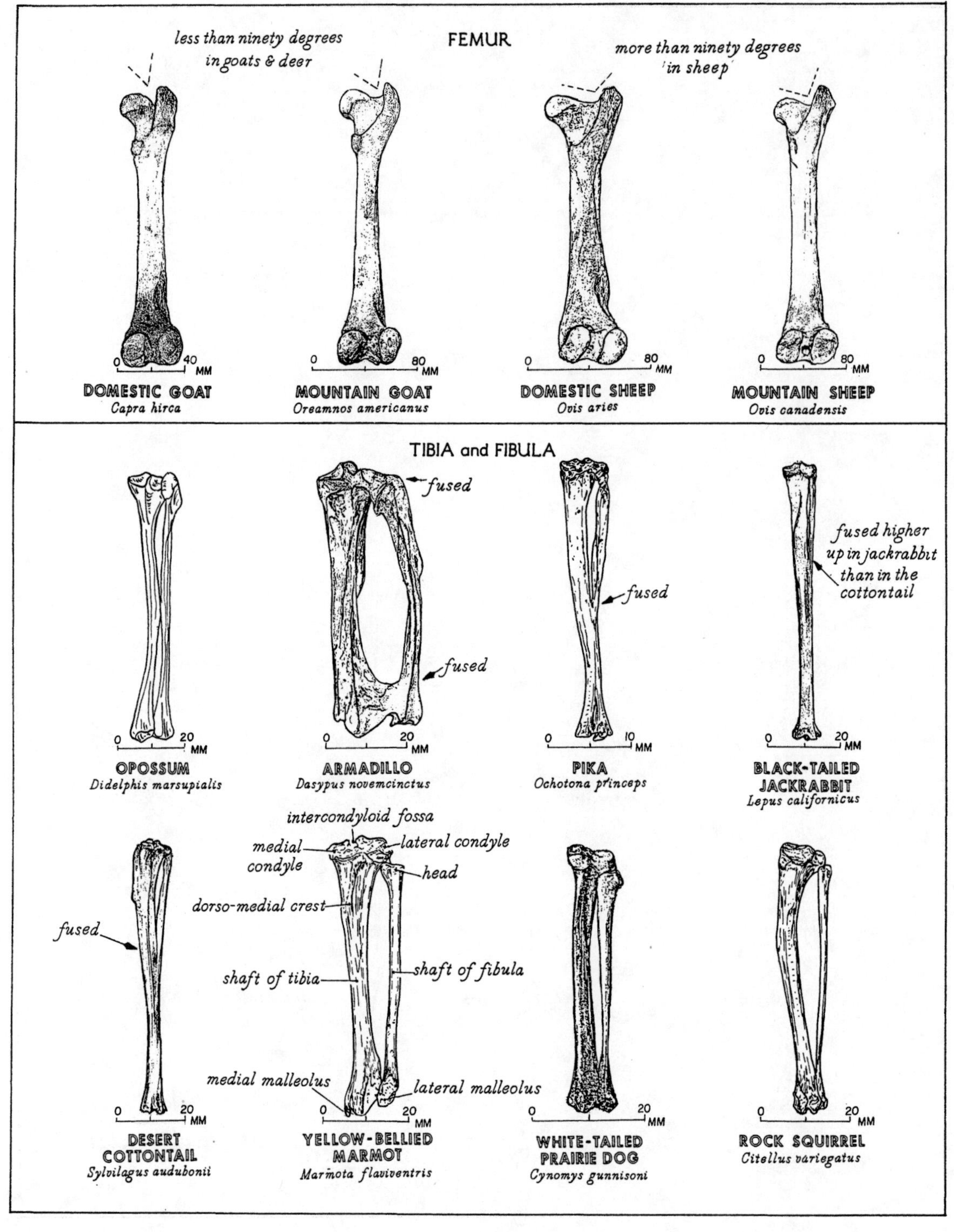
FEMUR
less than ninety degrees in goats & deer
more than ninety degrees in sheep
0 40 MM
DOMESTIC GOAT
Capra hirca
0 80 MM
MOUNTAIN GOAT
Oreamnos americanus
0 80 MM
DOMESTIC SHEEP
Ovis aries
0 80 MM
MOUNTAIN SHEEP
Ovis canadensis
TIBIA and FIBULA
0 20 MM
OPOSSUM
Didelphis marsupialis
fused
fused
0 20 MM
ARMADILLO
Dasypus novemcinctus
fused
0 10 MM
PIKA
Ochotona princeps
fused higher up in jackrabbit than in the cottontail
0 20 MM
BLACK-TAILED JACKRABBIT
Lepus californicus
fused
0 20 MM
DESERT COTTONTAIL
Sylvilagus audubonii
intercondyloid fossa
medial condyle
lateral condyle
head
dorso-medial crest
shaft of tibia
shaft of fibula
medial malleolus
lateral malleolus
0 20 MM
YELLOW-BELLIED MARMOT
Marmota flaviventris
0 20 MM
WHITE-TAILED PRAIRIE DOG
Cynomys gunnisoni
0 20 MM
ROCK SQUIRREL
Citellus variegatus

FIGURE 93

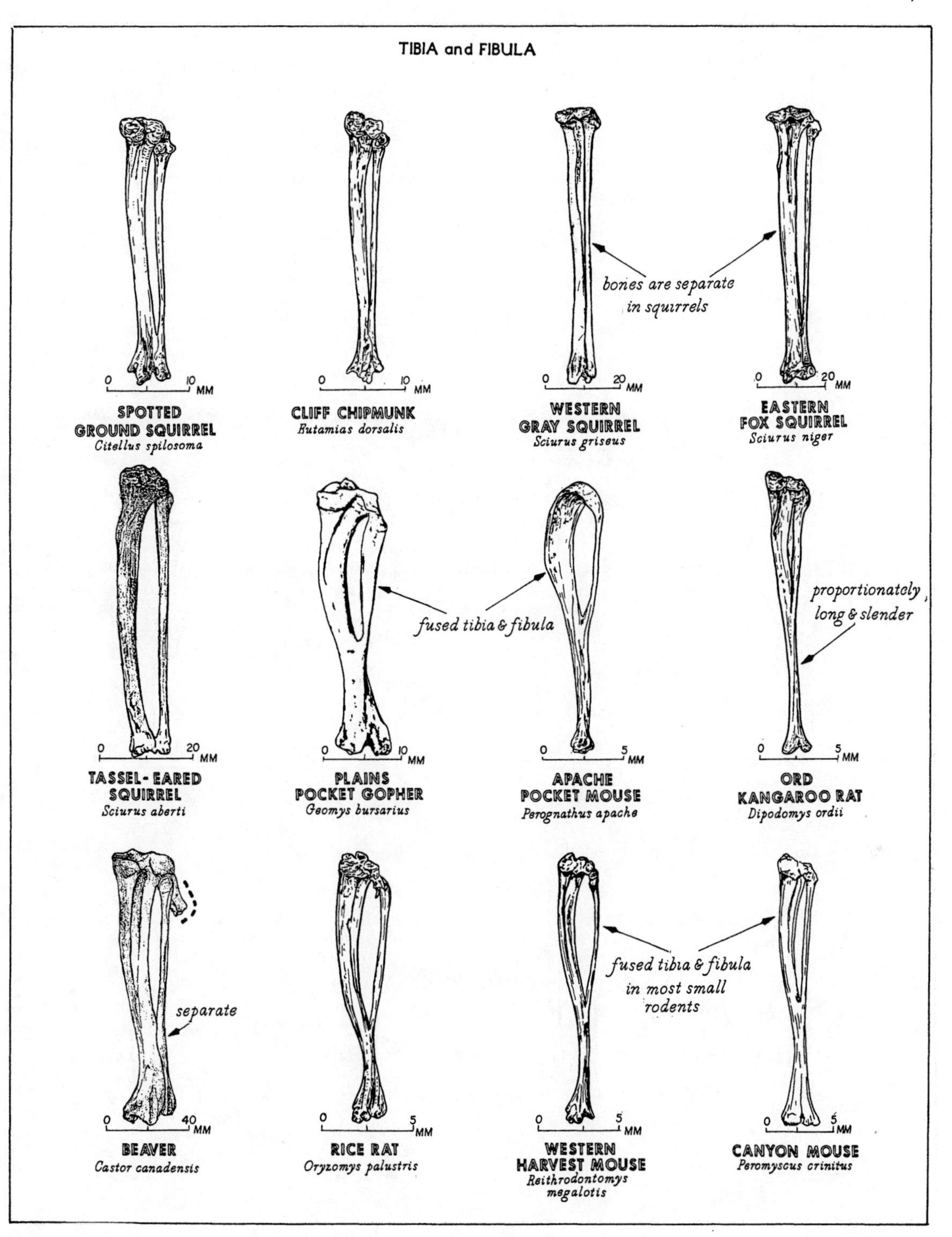
TIBIA and FIBULA
0 10 MM
SPOTTED GROUND SQUIRREL
Citellus spilosoma
0 10 MM
CLIFF CHIPMUNK
Eutamias dorsalis
bones are separate in squirrels
0 20 MM
WESTERN GRAY SQUIRREL
Sciurus griseus
0 20 MM
EASTERN FOX SQUIRREL
Sciurus niger
0 20 MM
TASSEL-EARED SQUIRREL
Sciurus aberti
fused tibia & fibula
0 10 MM
PLAINS POCKET GOPHER
Geomys bursarius
0 5 MM
APACHE POCKET MOUSE
Perognathus apache
proportionately long & slender
0 5 MM
ORD KANGAROO RAT
Dipodomys ordii
separate
0 40 MM
BEAVER
Castor canadensis
0 5 MM
RICE RAT
Oryzomys palustris
fused tibia & fibula in most small rodents
0 5 MM
WESTERN HARVEST MOUSE
Reithrodontomys megalotis
0 5 MM
CANYON MOUSE
Peromyscus crinitus

FIGURE 94

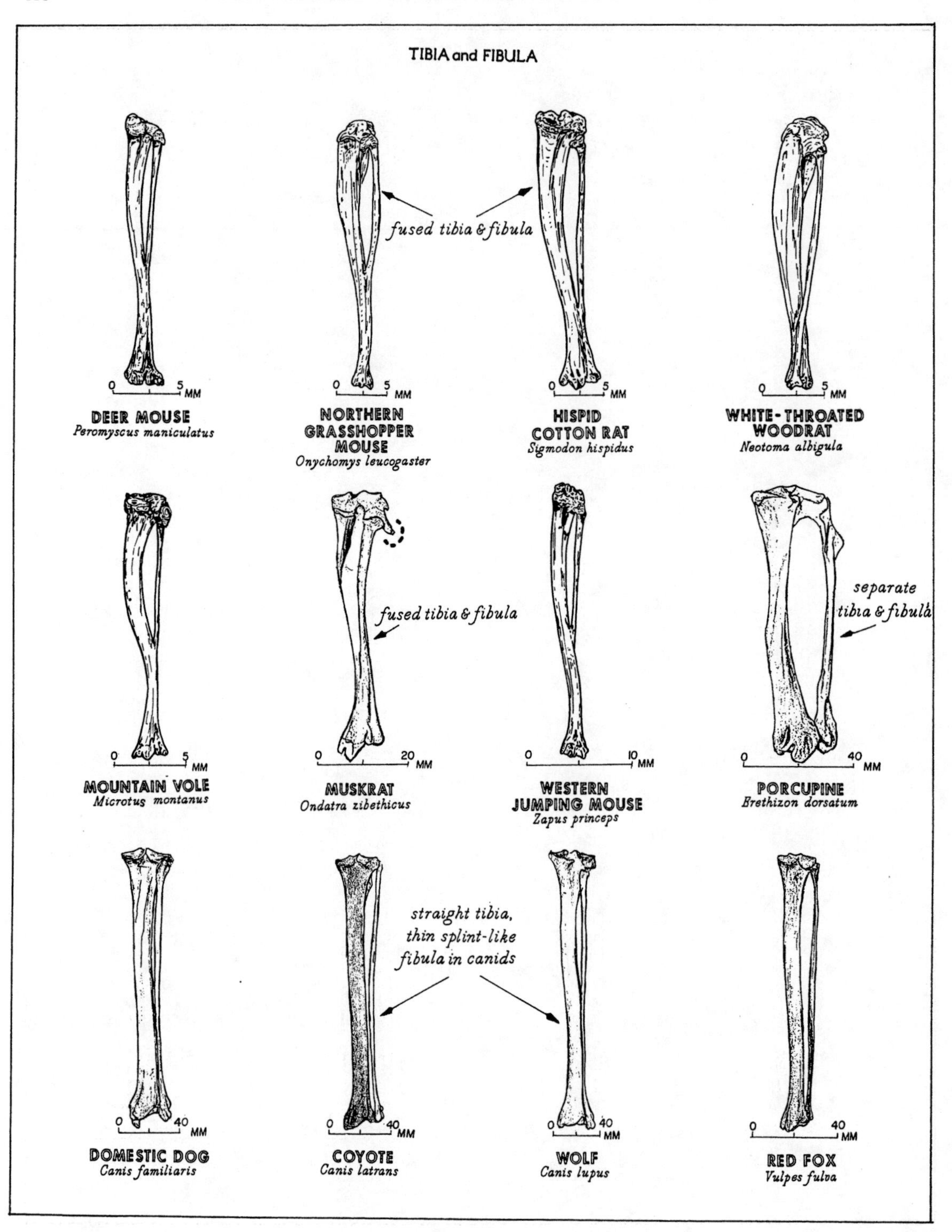
TIBIA and FIBULA
fused tibia & fibula
0 5 MM
DEER MOUSE
Peromyscus maniculatus
0 5 MM
NORTHERN GRASSHOPPER MOUSE
Onychomys leucogaster
0 5 MM
HISPID COTTON RAT
Sigmodon hispidus
0 5 MM
WHITE-THROATED WOODRAT
Neotoma albigula
fused tibia & fibula
separate tibia & fibula
0 5 MM
MOUNTAIN VOLE
Microtus montanus
0 20 MM
MUSKRAT
Ondatra zibethicus
0 10 MM
WESTERN JUMPING MOUSE
Zapus princeps
0 40 MM
PORCUPINE
Erethizon dorsatum
straight tibia, thin splint-like fibula in canids
0 40 MM
DOMESTIC DOG
Canis familiaris
0 40 MM
COYOTE
Canis latrans
0 40 MM
WOLF
Canis lupus
0 40 MM
RED FOX
Vulpes fulva

FIGURE 95

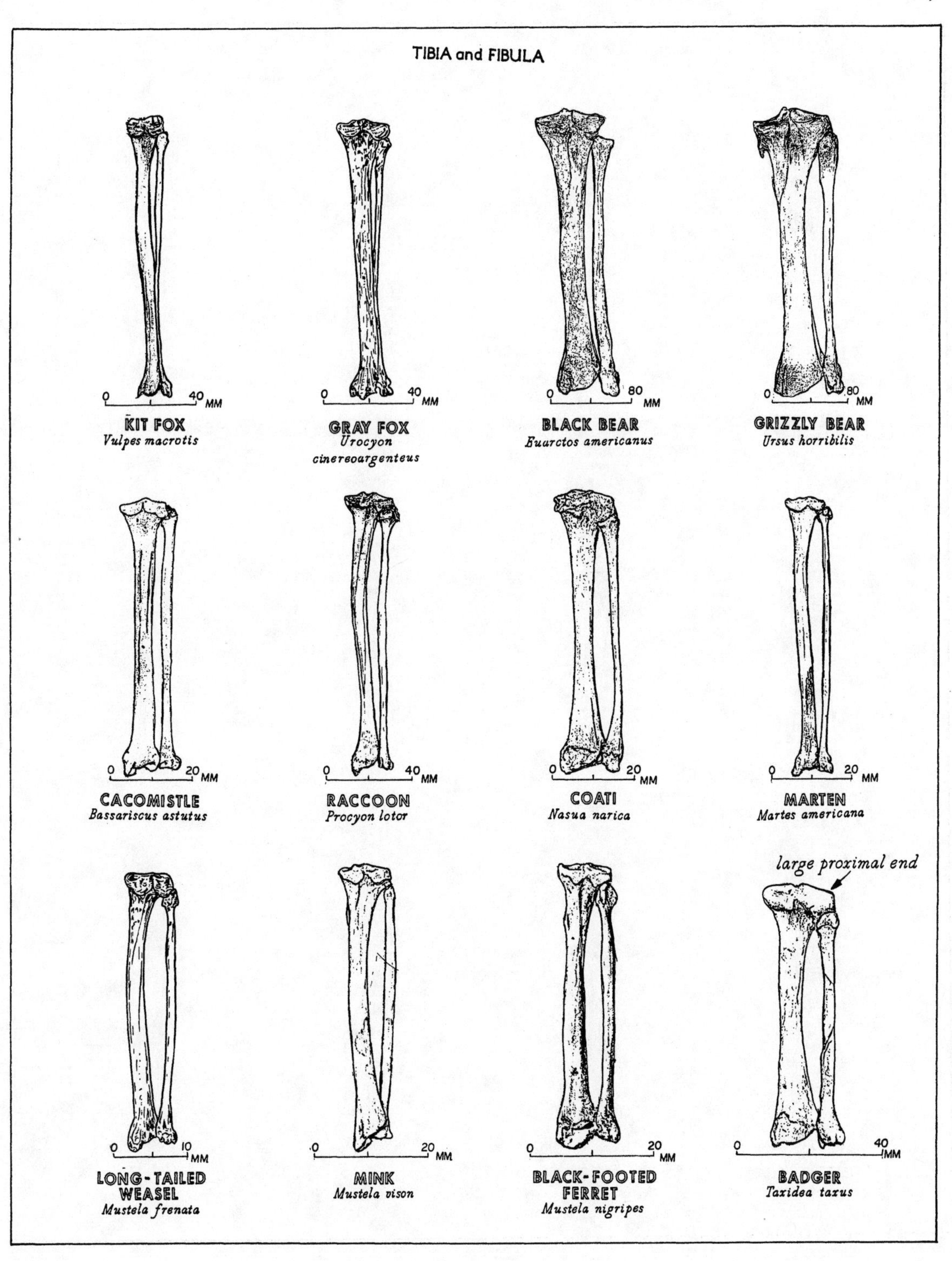
TIBIA and FIBULA
0 40 MM
KIT FOX
Vulpes macrotis
0 40 MM
GRAY FOX
Urocyon cinereoargenteus
0 80 MM
BLACK BEAR
Euarctos americanus
0 80 MM
GRIZZLY BEAR
Ursus horribilis
0 20 MM
CACOMISTLE
Bassariscus astutus
0 40 MM
RACCOON
Procyon lotor
0 20 MM
COATI
Nasua narica
0 20 MM
MARTEN
Martes americana
0 10 MM
LONG-TAILED WEASEL
Mustela frenata
0 20 MM
MINK
Mustela vison
0 20 MM
BLACK-FOOTED FERRET
Mustela nigripes
large proximal end
0 40 MM
BADGER
Taxidea taxus

FIGURE 96

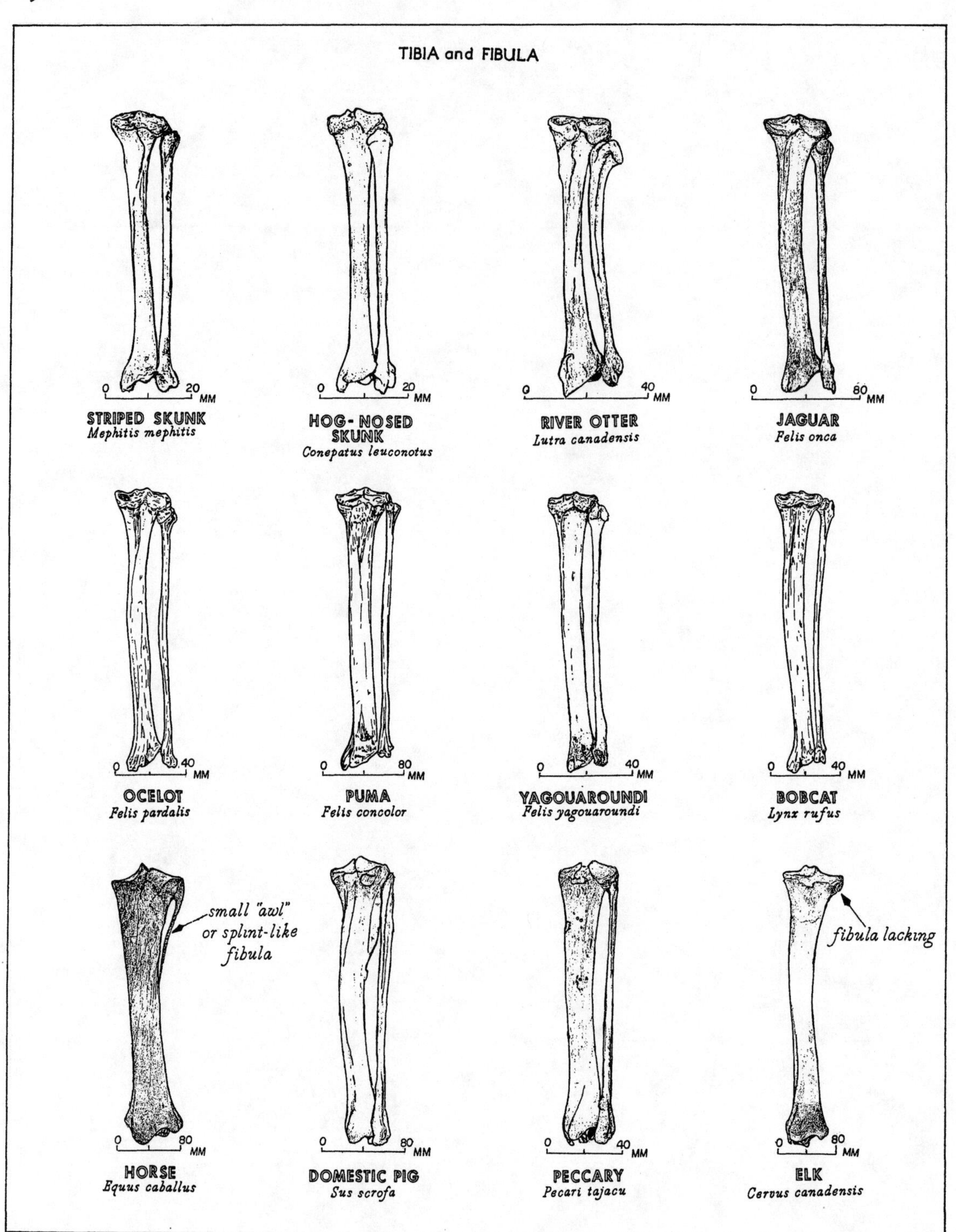
TIBIA and FIBULA
0 20 MM
STRIPED SKUNK
Mephitis mephitis
0 20 MM
HOG-NOSED SKUNK
Conepatus leuconotus
0 40 MM
RIVER OTTER
Lutra canadensis
0 80 MM
JAGUAR
Felis onca
0 40 MM
OCELOT
Felis pardalis
0 80 MM
PUMA
Felis concolor
0 40 MM
YAGOUAROUNDI
Felis yagouaroundi
0 40 MM
BOBCAT
Lynx rufus
small "awl" or splint-like fibula
0 80 MM
HORSE
Equus caballus
0 80 MM
DOMESTIC PIG
Sus scrofa
0 40 MM
PECCARY
Pecari tajacu
fibula lacking
0 80 MM
ELK
Cervus canadensis

FIGURE 97

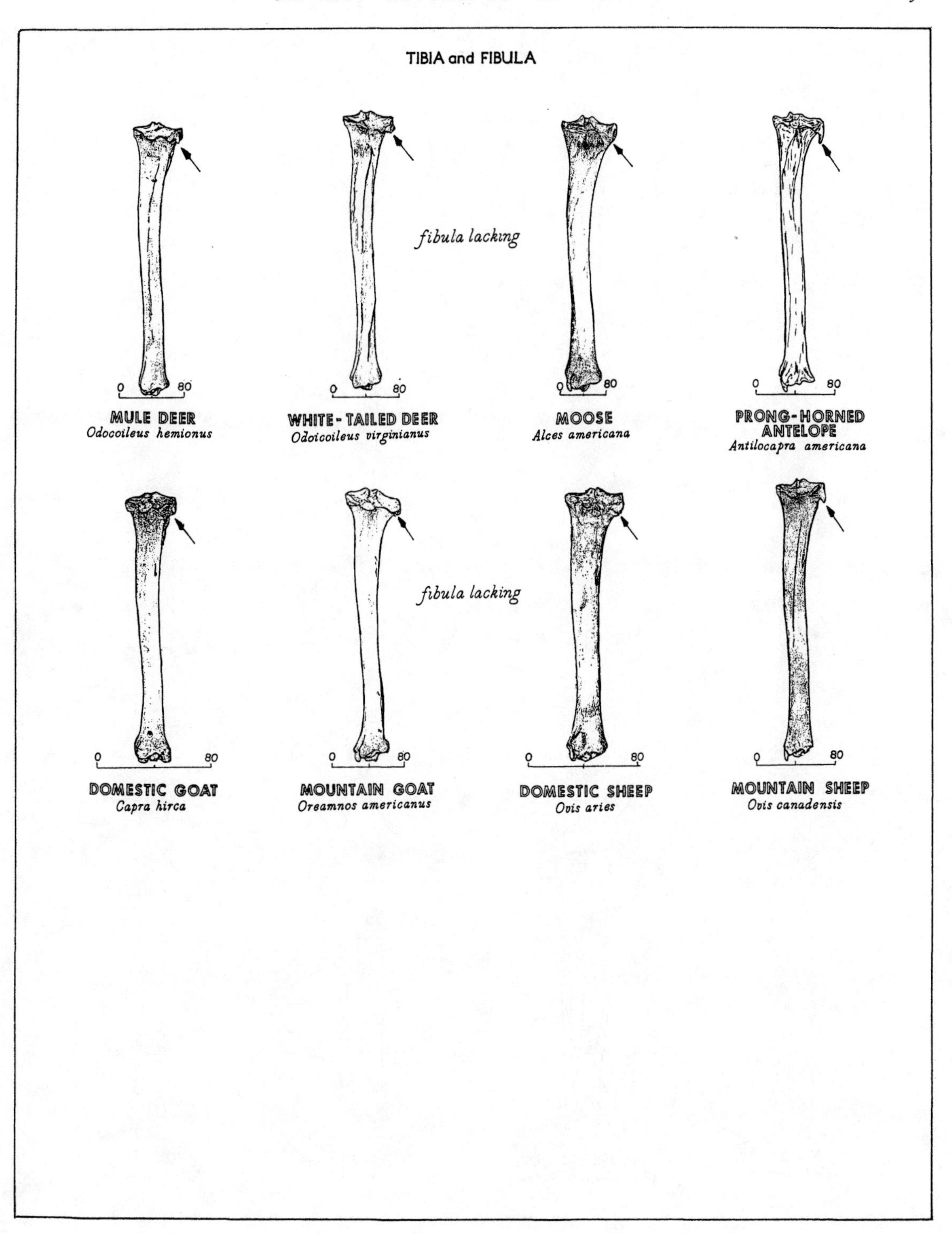
TIBIA and FIBULA
fibula lacking
0 80
MULE DEER
Odocoileus hemionus
0 80
WHITE-TAILED DEER
Odoicoileus virginianus
0 80
MOOSE
Alces americana
0 80
PRONG-HORNED ANTELOPE
Antilocapra americana
fibula lacking
0 80
DOMESTIC GOAT
Capra hirca
0 80
MOUNTAIN GOAT
Oreamnos americanus
0 80
DOMESTIC SHEEP
Ovis aries
0 80
MOUNTAIN SHEEP
Ovis canadensis

FIGURE 98

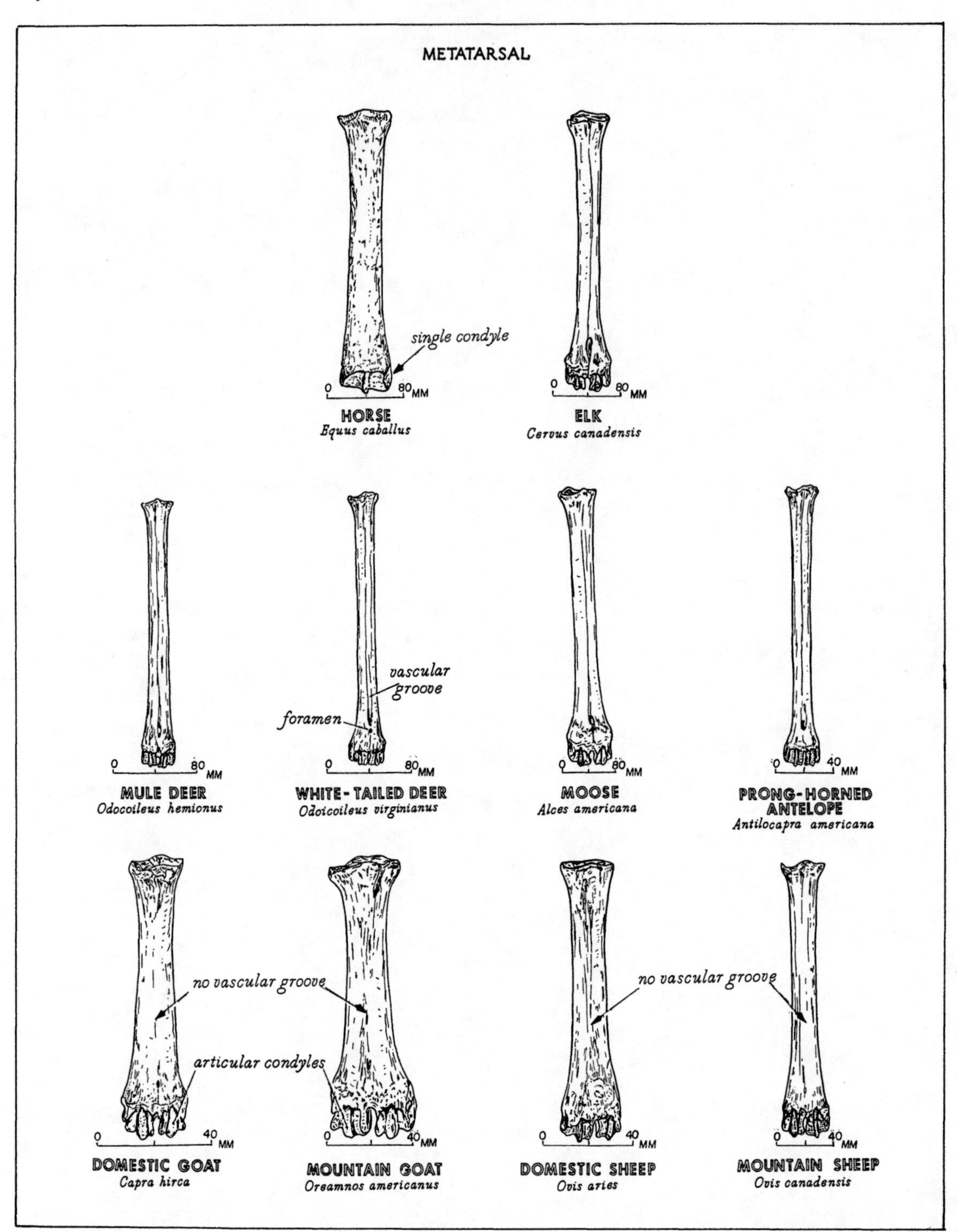
METATARSAL
single condyle
0 80 MM
HORSE
Equus caballus
0 80 MM
ELK
Cervus canadensis
0 80 MM
MULE DEER
Odocoileus hemionus
vascular groove
foramen
0 80 MM
WHITE-TAILED DEER
Odoicoileus virginianus
0 80 MM
MOOSE
Alces americana
0 40 MM
PRONG-HORNED ANTELOPE
Antilocapra americana
no vascular groove
articular condyles
0 40 MM
DOMESTIC GOAT
Capra hirca
0 40 MM
MOUNTAIN GOAT
Oreamnos americanus
no vascular groove
0 40 MM
DOMESTIC SHEEP
Ovis aries
0 40 MM
MOUNTAIN SHEEP
Ovis canadensis

FIGURE 99

FIGURES 100–116

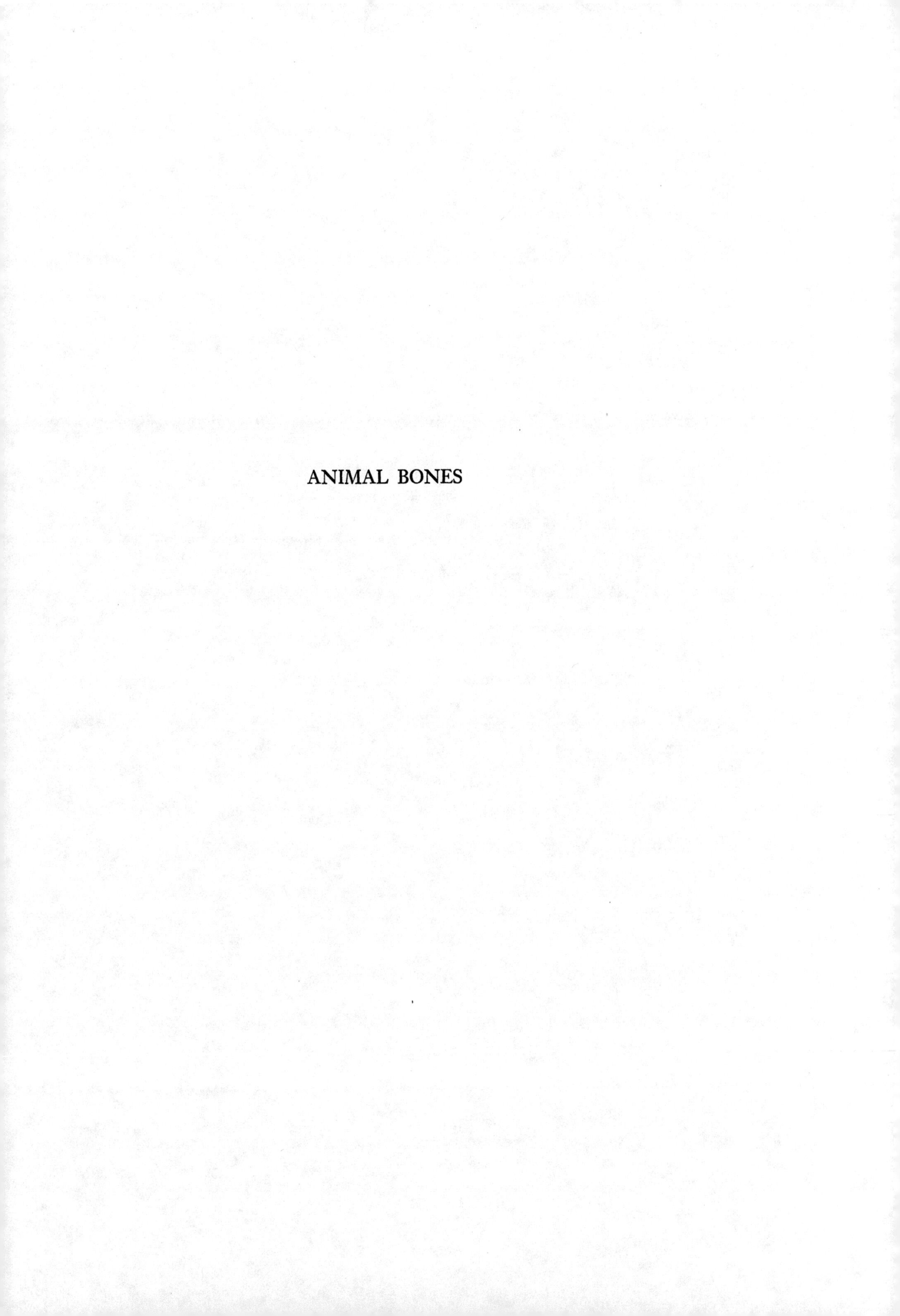

ANIMAL BONES

FIGURE 100

Animal bones modified for use as tools

A. Unaltered deer radius and ulna.
B. Awl or flaking tool made from deer ulna. From a mound in Hillsborough County, Florida.
C. Unaltered deer metatarsal.
D. Awl or fish spear made from deer metatarsal. From the Wakulla River, Florida.
E. Unaltered deer humerus.
F. Scraper made from deer humerus. From Pueblo Bonito, New Mexico.
G. Scraper made from deer humerus, inlaid with shell, jet, and turquoise. From Pueblo Bonito, New Mexico.

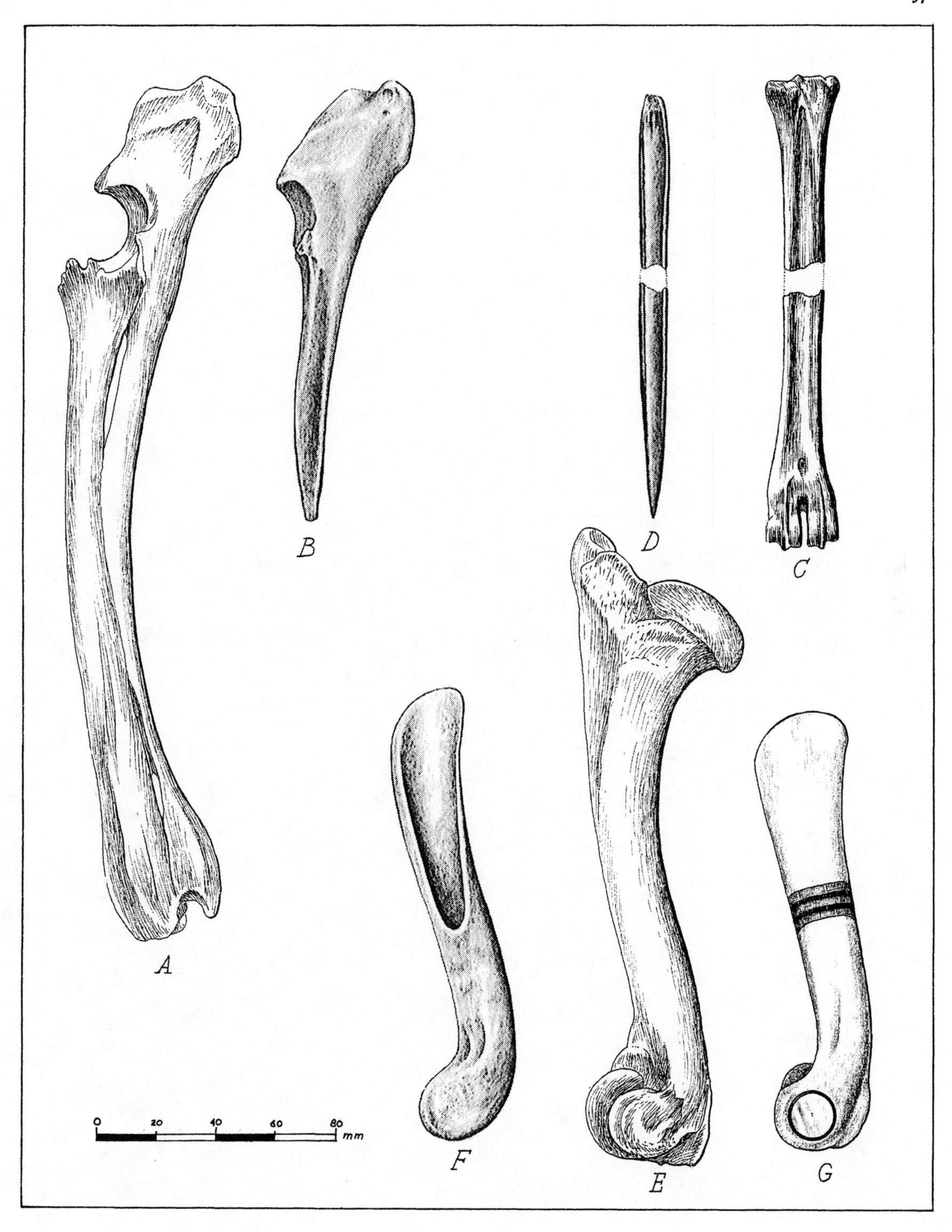
A
B
C
D
E
F
G
0
20
40
60
80
mm

BODY FORMS OF ANIMALS

FIGURES 101 TO 108

Body forms of animals included in this study.

The form and size of the mammals discussed in this paper are included in the belief that more interest is shown by the student in identifying isolated bones if he has some idea of the animal's appearance in life. In some isolated instances, native carvings and sculpture depict animal forms that can be compared with these plates.

CACOMISTLE
Bassariscus astutus
OPOSSUM
Didelphis marsupialis
PORCUPINE
Erethizon dorsatum
COATI
Nasua narica
ARMADILLO
Dasypus novemcinctus
DESERT COTTONTAIL
Sylvilagus audubonii
STRIPED SKUNK
Mephitis mephitis
RACCOON
Procyon lotor
BLACK-TAILED JACKRABBIT
Lepus californicus
DOMESTIC PIG
Sus scrofa
PECCARY
Pecari tajacu
A.R. Janson

FIGURE 101

SPOTTED GROUND SQUIRREL
Citellus spilosoma
ROCK SQUIRREL
Citellus variegatus
PIKA
Ochotona princeps
EASTERN FOX SQUIRREL
Sciurus niger
WESTERN GRAY SQUIRREL
Sciurus griseus
MUSKRAT
Ondatra zibethicus
WHITE-TAILED PRAIRIE DOG
Cynomys gunnisoni
TASSEL-EARED SQUIRREL
Sciurus aberti
YELLOW-BELLIED MARMOT
Marmota flaviventris
BEAVER
Castor canadensis
A.R. Janson

FIGURE 102

ORD KANGAROO RAT
Dipodomys ordii
WESTERN HARVEST MOUSE
Reithrodontomys megalotis
APACHE POCKET MOUSE
Perognathus apache
DEER MOUSE
Peromyscus maniculatus
CANYON MOUSE
Peromyscus crinitus
MOUNTAIN VOLE
Microtus montanus
WESTERN JUMPING MOUSE
Zapus princeps
CLIFF CHIPMUNK
Eutamias dorsalis
WHITE-THROATED WOOD RAT
Neotoma albigula
HISPID COTTON RAT
Sigmodon hispidus
PLAINS POCKET GOPHER
Geomys bursarius
NORTHERN GRASSHOPPER MOUSE
Onychomys leucogaster
A.R. Janson

FIGURE 103

INDIAN DOG
Canis familiaris
WOLF
Canis lupus
COYOTE
Canis latrans
KIT FOX
Vulpes macrotis
GRAY FOX
Urocyon cinereoargenteus
BADGER
Taxidea taxus
BOBCAT
Lynx rufus
JAGUAR
Felis onca
PUMA
Felis concolor
BLACK-FOOTED FERRET
Mustela nigripes
A.B. Janson

FIGURE 104

LONG-TAILED WEASEL
Mustela frenata
MINK
Mustela vison
RIVER OTTER
Lutra canadensis
MARTEN
Martes americana
YAGOUAROUNDI
Felis yagouaroundi
OCELOT
Felis pardalis
BLACK BEAR
Euarctos americanus
GRIZZLY BEAR
Ursus horribilis
A.R. Janson

FIGURE 105

HORSE
Equus caballus
BURRO
Equus asinus
MULE
Equus asinus ♂ Equus caballus ♀
A.R. Janson

FIGURE 106

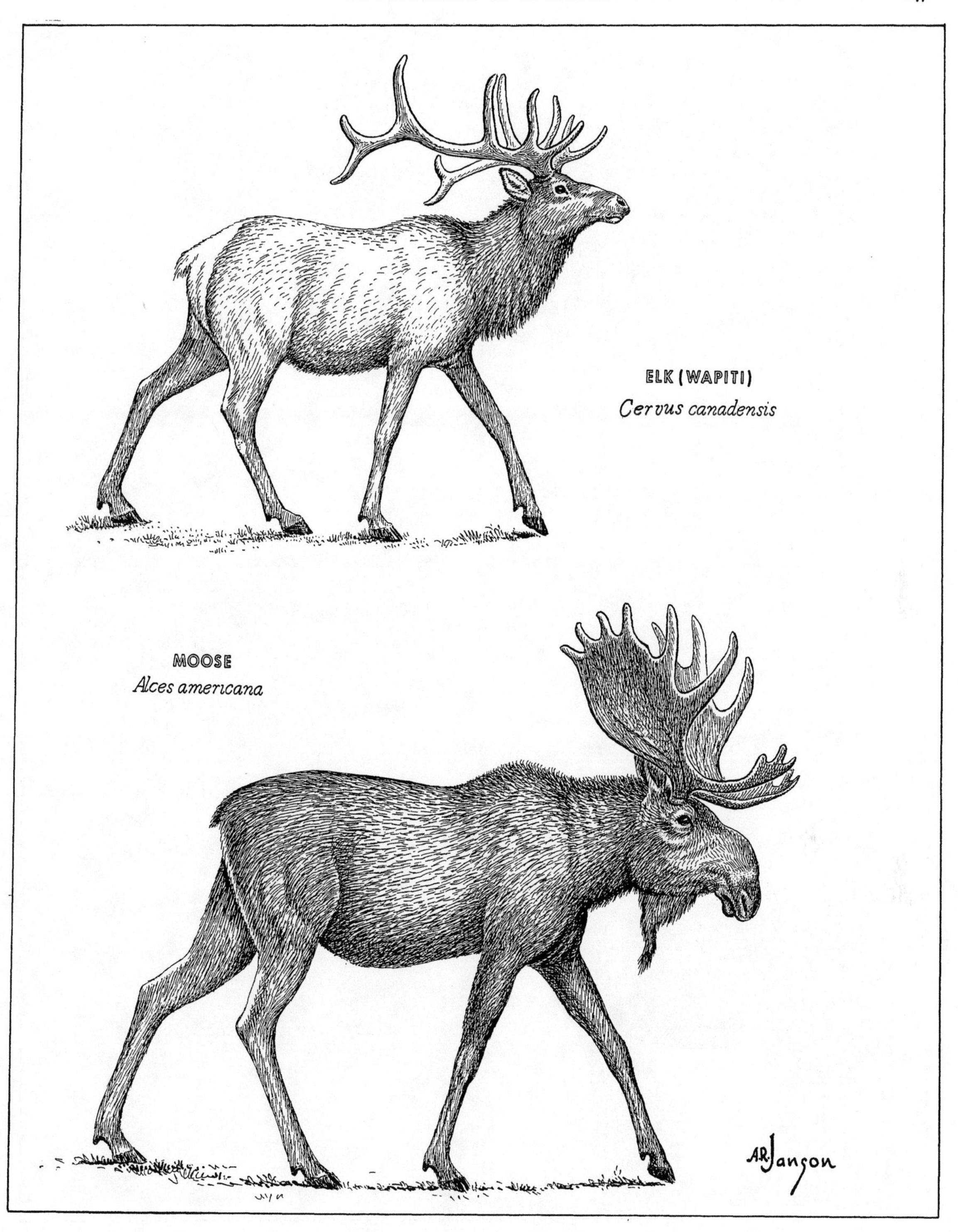
ELK (WAPITI)
Cervus canadensis
MOOSE
Alces americana
A.R. Janson

FIGURE 107

DOMESTIC SHEEP
Ovis aries
MOUNTAIN SHEEP
Ovis canadensis
DOMESTIC GOAT
Capra hirca
MOUNTAIN GOAT
Oreamnos americanus
WHITE-TAILED DEER
Odocoileus virginianus
PRONG-HORNED ANTELOPE
Antilocapra americana
MULE DEER
Odocoileus hemionus
A.B. Janson

FIGURE 108

DISTRIBUTION MAPS OF ANIMALS

Figures 109 to 116

Distribution maps of animals discussed in text.

These charts are approximate, based on known records of the animals concerned. In some instances more careful analysis of osteological material in archæological collections will alter the limits as here indicated.

Since this study is primarily concerned with mammals found in the southern portion of the United States, the distribution patterns of those forms which extend beyond the northern and southern boundaries of the United States are not shown in their entirety.

The skeletal remains of the pig *Sus scrofa* and other domestic forms may be encountered in any area occupied by civilized man. The distribution maps of domestic animals do not necessarily imply that these forms lived in all of the areas within this pattern but rather that their bones may occur as intrusive burials (transported by flood wash, rodents, etc.) in older sites within these areas, after having been brought to the vicinity by modern man.

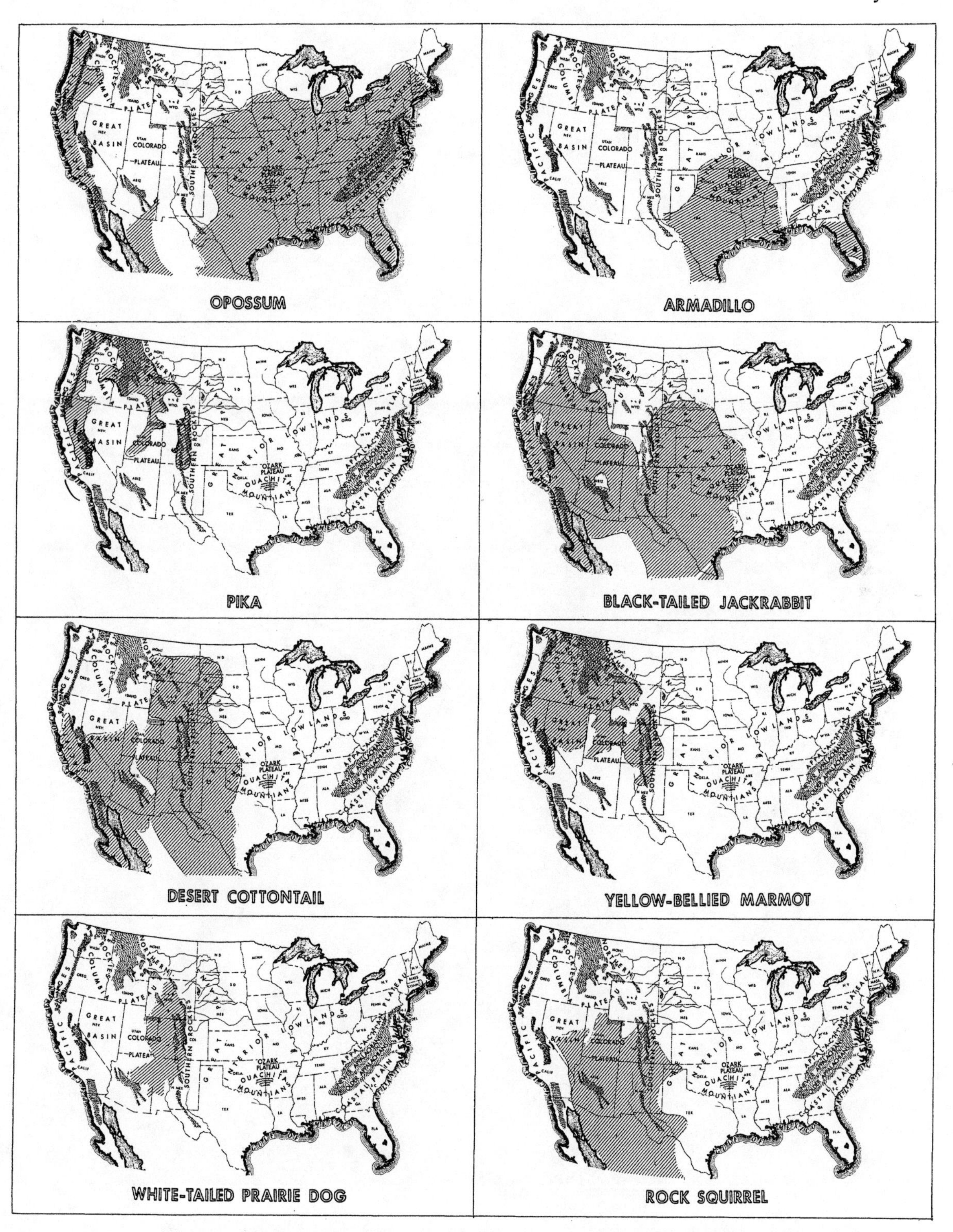
OPOSSUM
ARMADILLO
PIKA
BLACK-TAILED JACKRABBIT
DESERT COTTONTAIL
YELLOW-BELLIED MARMOT
WHITE-TAILED PRAIRIE DOG
ROCK SQUIRREL

FIGURE 109

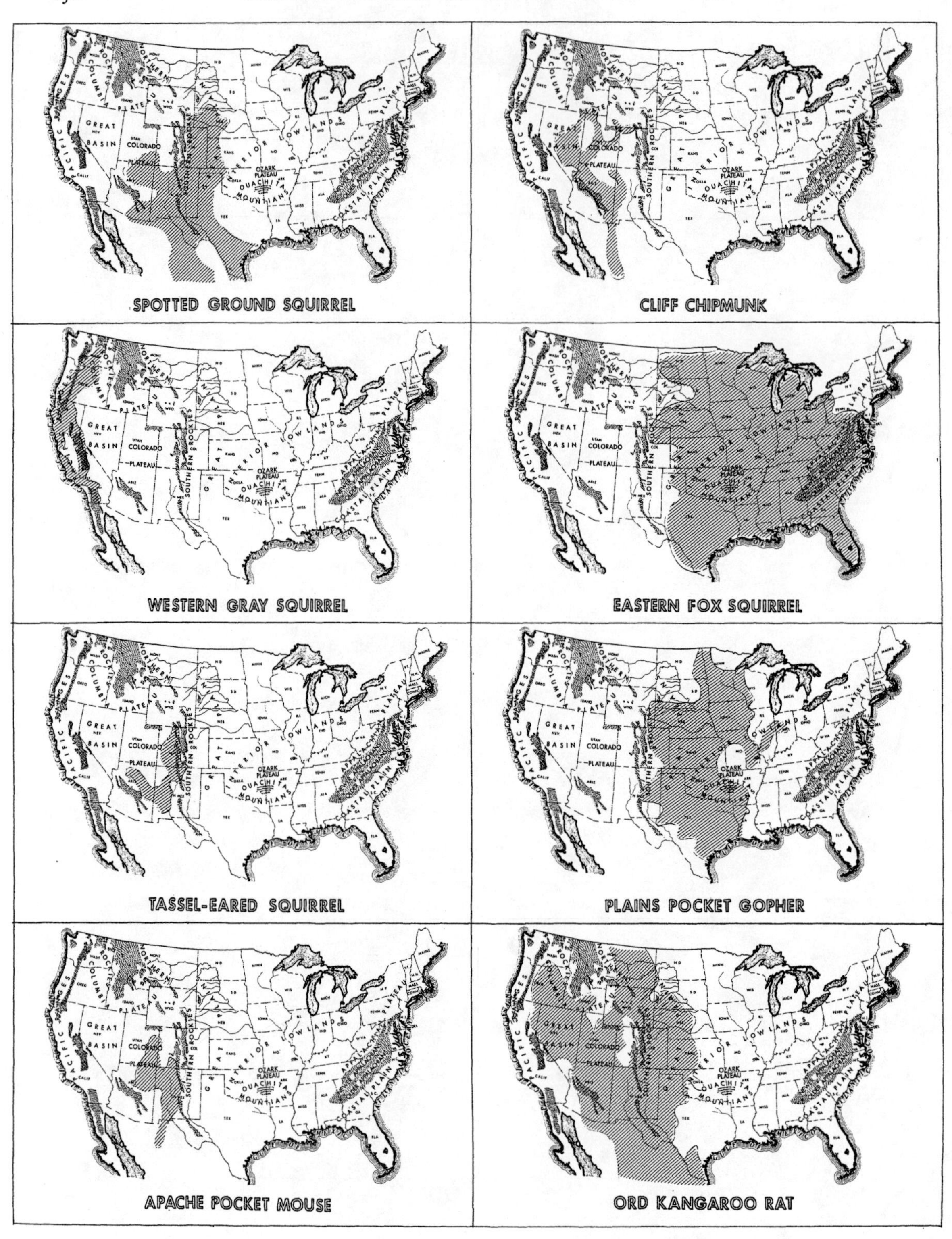
SPOTTED GROUND SQUIRREL
CLIFF CHIPMUNK
WESTERN GRAY SQUIRREL
EASTERN FOX SQUIRREL
TASSEL-EARED SQUIRREL
PLAINS POCKET GOPHER
APACHE POCKET MOUSE
ORD KANGAROO RAT

FIGURE 110

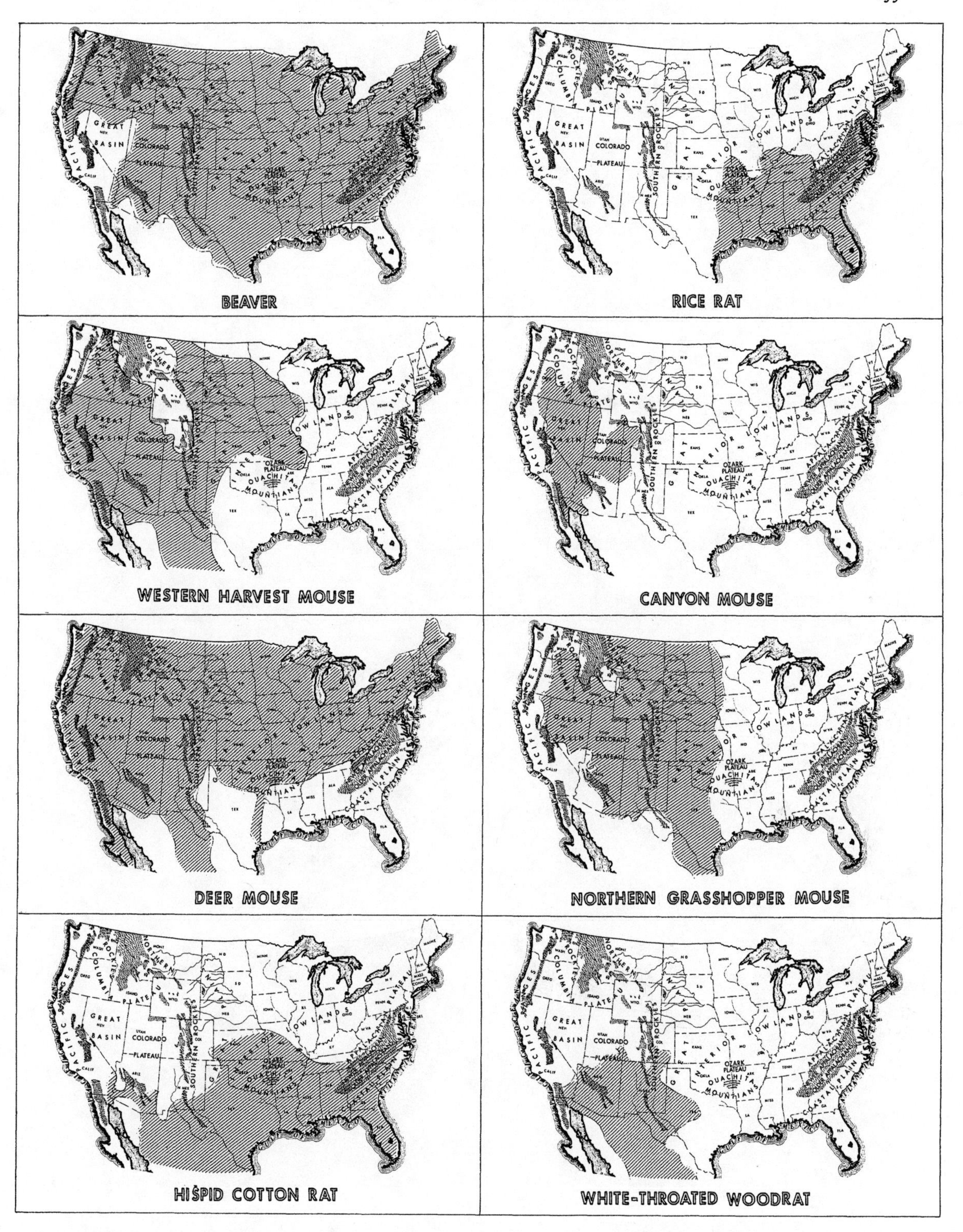
BEAVER
RICE RAT
WESTERN HARVEST MOUSE
CANYON MOUSE
DEER MOUSE
NORTHERN GRASSHOPPER MOUSE
HISPID COTTON RAT
WHITE-THROATED WOODRAT

FIGURE 111

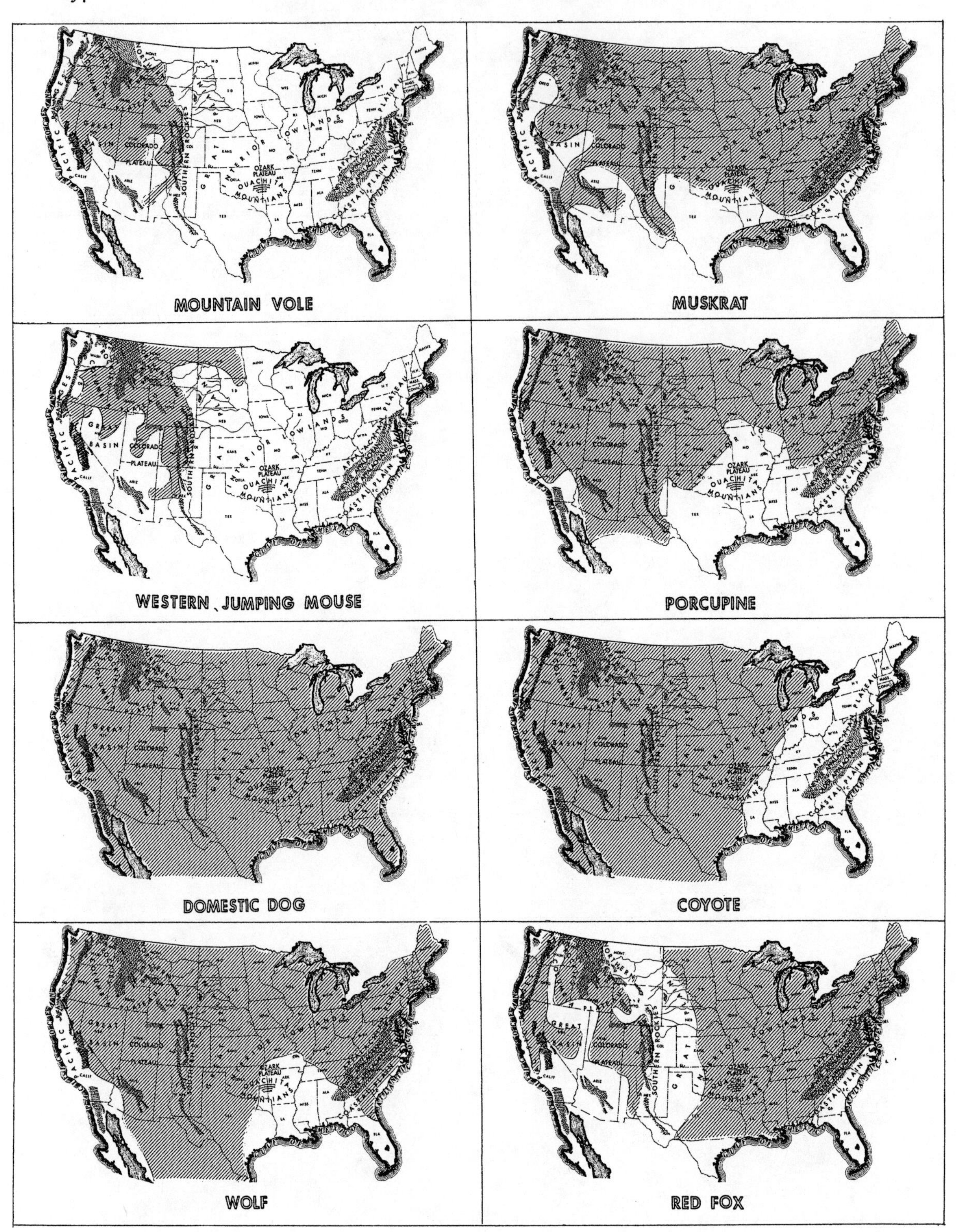
MOUNTAIN VOLE
MUSKRAT
WESTERN JUMPING MOUSE
PORCUPINE
DOMESTIC DOG
COYOTE
WOLF
RED FOX

FIGURE 112

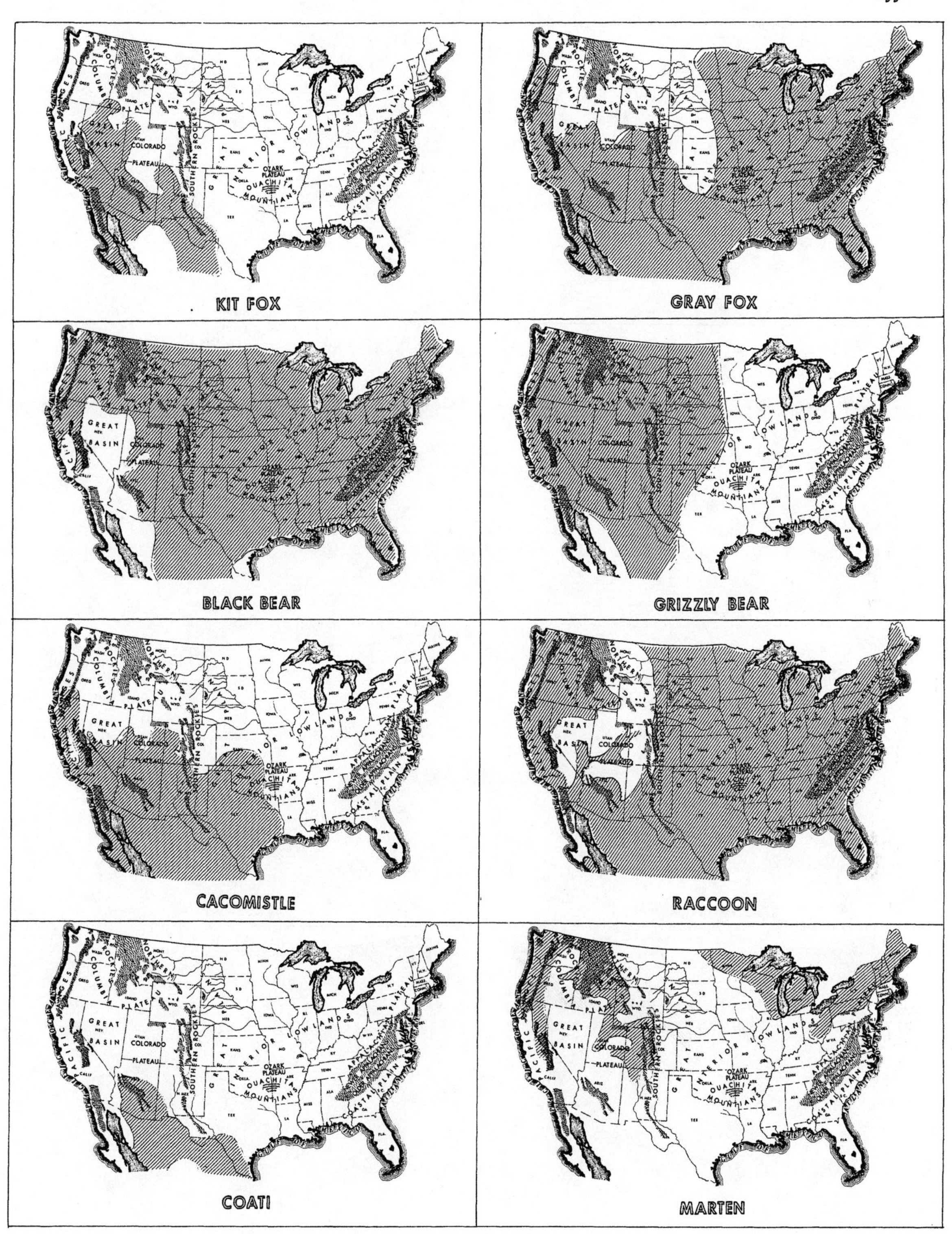
KIT FOX
GRAY FOX
BLACK BEAR
GRIZZLY BEAR
CACOMISTLE
RACCOON
COATI
MARTEN

FIGURE 113

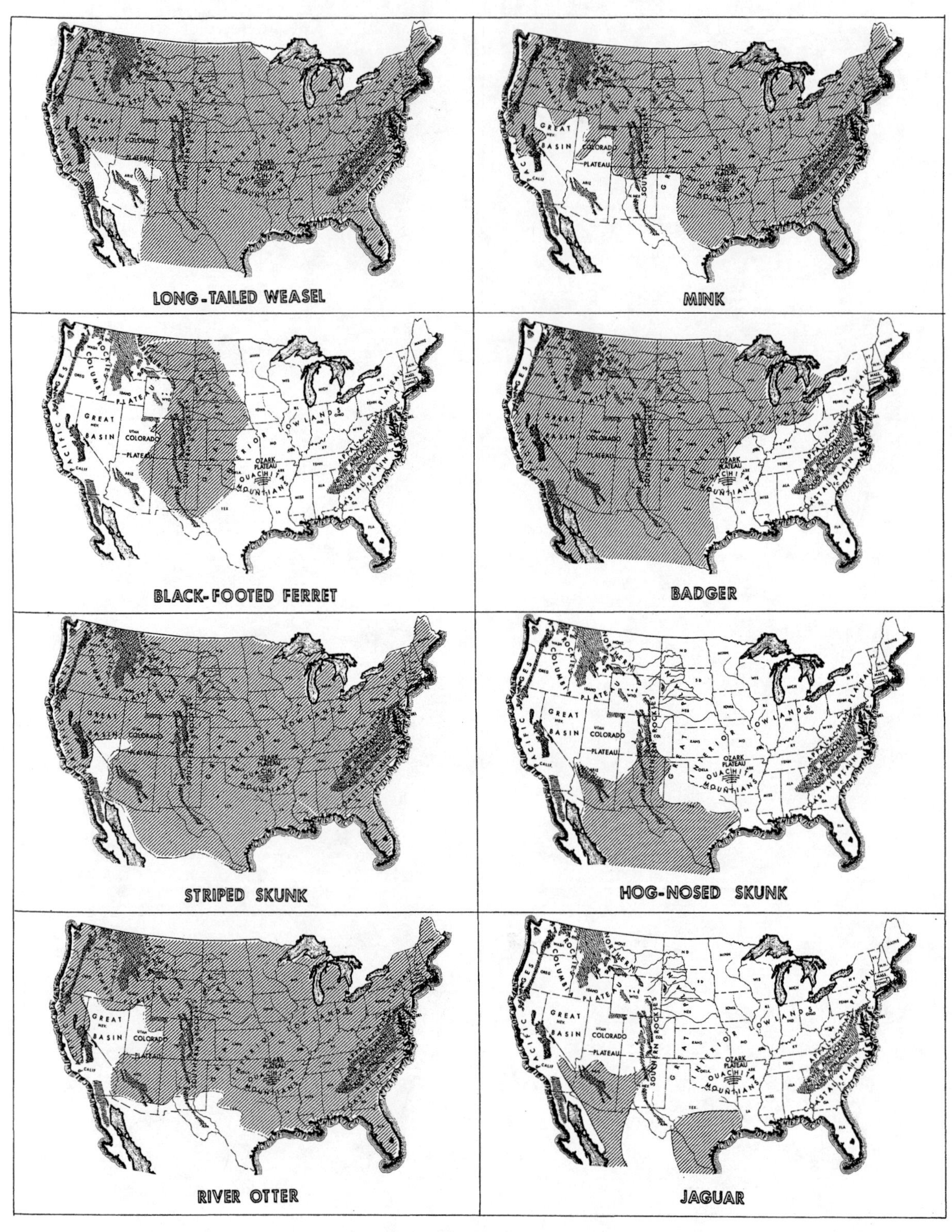
LONG-TAILED WEASEL
MINK
BLACK-FOOTED FERRET
BADGER
STRIPED SKUNK
HOG-NOSED SKUNK
RIVER OTTER
JAGUAR

FIGURE 114

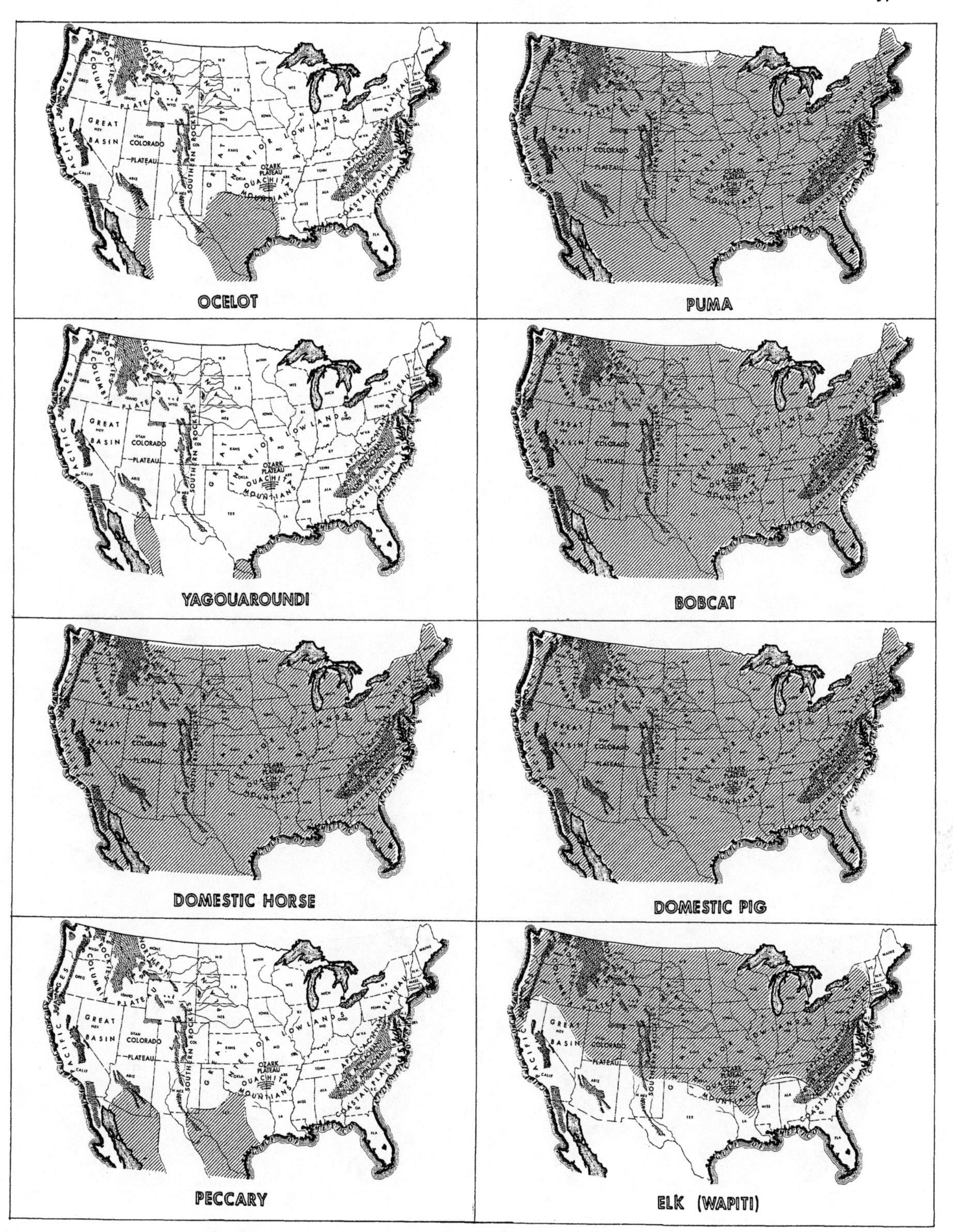
OCELOT
PUMA
YAGOUAROUNDI
BOBCAT
DOMESTIC HORSE
DOMESTIC PIG
PECCARY
ELK (WAPITI)

FIGURE 115

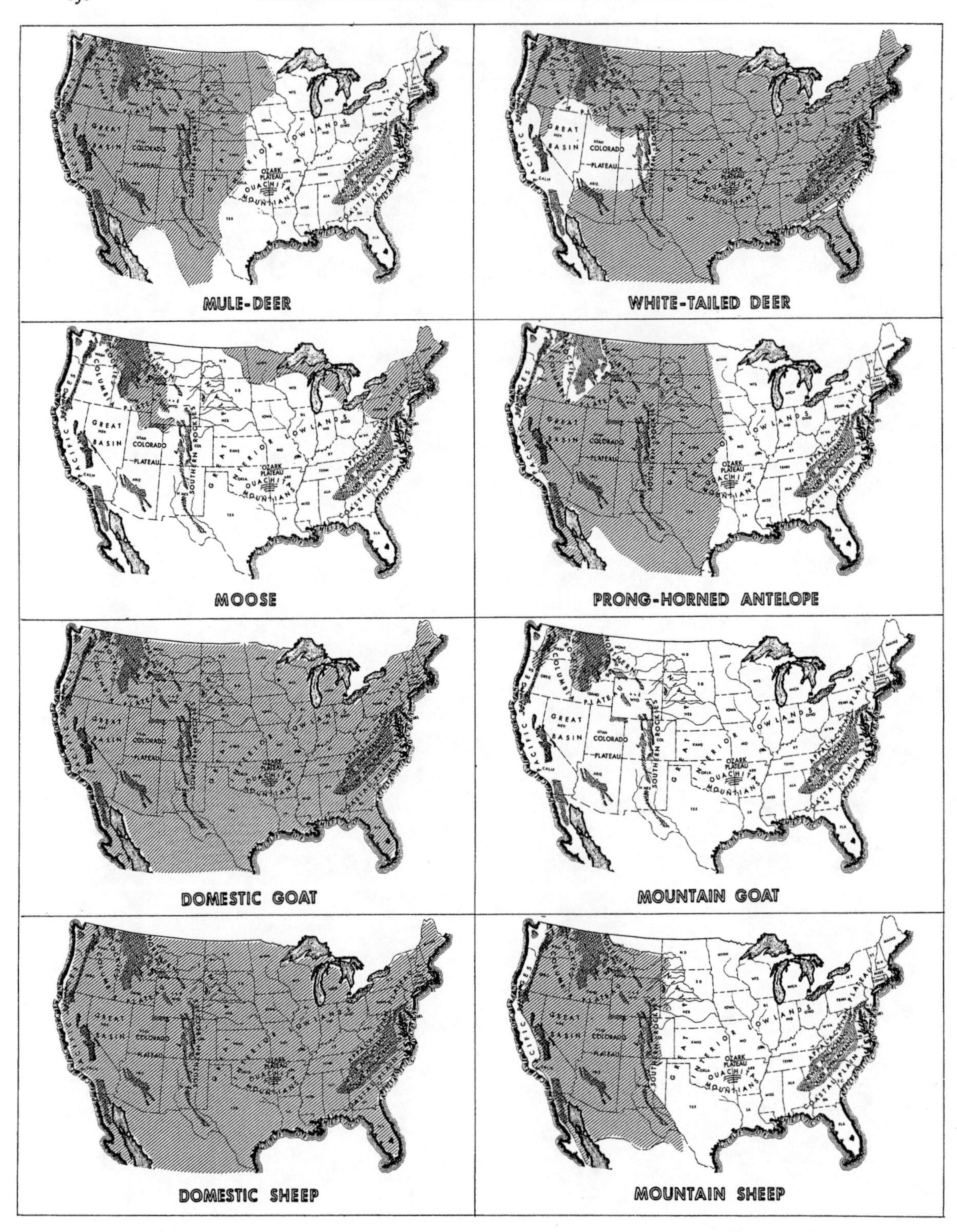
MULE-DEER
WHITE-TAILED DEER
MOOSE
PRONG-HORNED ANTELOPE
DOMESTIC GOAT
MOUNTAIN GOAT
DOMESTIC SHEEP
MOUNTAIN SHEEP
GREAT BASIN
COLORADO PLATEAU
SOUTHERN ROCKIES
OZARK PLATEAU
OUACHITA MOUNTAINS
INTERIOR LOWLANDS
COASTAL PLAIN

FIGURE 116

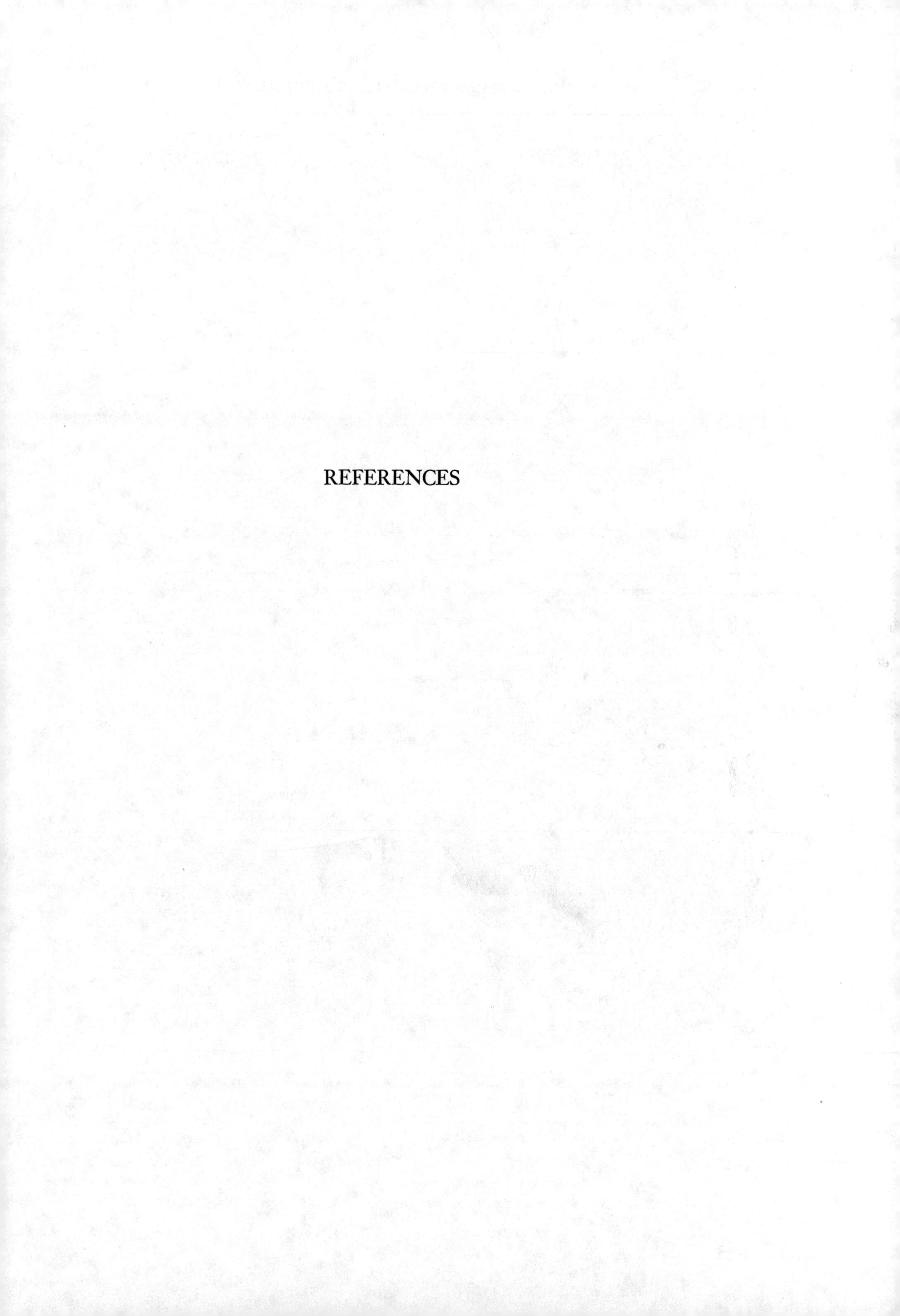

REFERENCES

REFERENCES

Blair, W. F., Blair, A. P., Brodkorb, P., Cagle, F. R., and Moore, G. A.
1957. Vertebrates of the United States. 819 pages. New York.

Booth, E. S.
1950. How to know the mammals. 206 pages. Dubuque, Iowa.

Brainerd, G. W.
1939. An illustrated key for the identification of mammal bones. *Ohio State Archaeological and Historical Quarterly*, vol. 48, no. 4, pp. 324–28. Columbus, Ohio.

Brown, G. H.
1952. Illustrated skull key to the recent land mammals of Virginia. *Virginia Cooperative Wildlife, Research Unit*, 75 pages. Blacksburg, Virginia.

Burt, W. H.
1960. Bacula of North American mammals. *Museum of Zoology, University of Michigan, Miscellaneous Publication* 113, 75 pages. Ann Arbor, Michigan.

Burt, W. H., and Grossenheider, R. P.
1952. A field guide to the mammals. 200 pages. Boston.

Cockrum, E. L.
1960. The recent mammals of Arizona. 276 pages. Tucson.
1962. Introduction to mammalogy. 455 pages. New York.

Colbert, E. H.
1955. Evolution of the vertebrates. 479 pages. New York.

Cornwall, I. W.
1956. Bones for the archaeologist. 255 pages. New York.

Davis, W. B.
1960. The mammals of Texas. *Texas Game & Fish Comm.*, 252 pages. Austin.

Ewart, J. C.
1904. The multiple origin of horses and ponies. *Smithsonian Institution, Report* no. 430, pt. 1, pp. 437–55. Washington, D.C.

Flower, W. H.
1876. An introduction to the osteology of the mammalia. 339 pages. London.

Flower, W. H., and Lydekker, R.
1891. An introduction to the study of mammals living and extinct. 763 pages. London.

Gier, H. T.
1957. Coyotes in Kansas. *Agricultural Experiment Station, Kansas State College, Bulletin* 393, 95 pages. Manhattan.

Glass, B. P.
1951. A key to the skulls of North American mammals. 54 pages. Minneapolis.

Greene, E. C.
1959. Anatomy of the rat. *Transactions of the American Philosophical Society, Philadelphia*, n.s., vol. XXXVII (reprint), 370 pages. New York.

Hall, E. R.
1946. Mammals of Nevada. 710 pages. Berkeley, California.
1955. Handbook of mammals of Kansas. *University of Kansas*, 303 pages. Lawrence.

Hall, E. R., and Kelson, K. R.
1959. The mammals of North America. 2 vols., 1,083 pages. New York.

Hamilton, W. J.
1943. The mammals of Eastern United States. 432 pages. Ithaca, New York.

Heizer, R. F., and Cook, S. F.
1960. The application of quantitive methods in archaeology. *Viking Fund Publication in Anthropology*, no. 28, 358 pages. Chicago.

Hildebrand, M.
1954. Comparative morphology of the body skeleton in recent canidae. *University of California, Publication in Zoology*, vol. 52, no. 5, pp. 399–470. Berkeley, California.
1955. Skeletal differences between deer, sheep, and goats. *California Fish & Game Comm.*, pp. 327–46. Davis.

Hoffmeister, D. F.
1962. The kinds of deer, odocoileus, in Arizona. *The American Midland Naturalist*, vol. 67, no. 1, January, pp. 45–64. Notre Dame.

Hoffmeister, D. F., and Mohr, C. O.
1957. Fieldbook of Illinois mammals. *Illinois Natural History Survey*, manual 4, 233 pages. Urbana.

HOWARD, W. E.
1949. A means to distinguish skulls of coyotes and domestic dogs. *Journal of Mammalogy*, vol. 30, no. 2, May, pp. 169–71. Salt Lake City.

HUE, E.
1907. Musée ostéologique. 2 vols., 186 plates. Paris.

JACKSON, H. H.
1961. Mammals of Wisconsin. 504 pages. Madison, Wisconsin.

LAWRENCE, B.
1951. Post-cranial skeletal characters of deer, pronghorn, and sheep-goat with notes on bos and bison. *Peabody Museum Papers, Harvard University*, vol XXXV, no. 3, 43 pages. Cambridge.

LEOPOLD, A. S.
1959. Wildlife of Mexico. 568 pages. Berkeley, California.

LOOMIS, F. B.
1926. The evolution of the horse. 233 pages. Boston.

LULL, R. S.
1931. The evolution of the horse family. *Yale Peabody Museum*, spec. guide no. 1, 29 pages. New Haven.

MCKENNA, M. C.
1962. Eupetaurus and the living Petauristine Sciuridas. *American Museum Novitates*, no. 2104, October, 37 pages. New York.

MILLER, M.
1952. Guide to the dissection of the dog. 369 pages. Ithaca, New York.

MILLER, G. S., JR., AND KELLOGG, R.
1955. List of North American recent mammals. *U.S. National Museum, Bulletin* 205, 954 pages. Washington, D.C.

MOORE, J. C.
1959. Relationships among living squirrels of the Sciurinae. *American Museum of Natural History, Bulletin*, vol. 118, art. 4, pp. 159–206. New York.

OLSEN, S. J.
1960. Post-cranial skeletal characters of bison and bos. *Peabody Museum Papers, Harvard University*, vol. XXXV, no. 4, 15 pages. Cambridge.
1961a. The relative value of fragmentary mammalian remains. *American Antiquity*, vol. 26, no. 4, April, pp. 538–40. Salt Lake City.
1961b. Problems of mammal skull identification due to age differences in the dentition. *American Antiquity*, vol. 27, no. 2, October, pp. 231–34. Salt Lake City.

POPE, G. W.
1934. Determining the age of farm animals by their teeth. *Farmers Bulletin* no. 1721, *U.S. Department of Agriculture*, 13 pages. Washington, D.C.

QUINN, J. H.
1957. Pleistocene Equidae of Texas. *Bureau of Economic Geology, University of Texas*, 51 pages. Austin.

RAY, C. E.
1957. Pre-Columbian horses from Yucatan. *Journal of Mammalogy*, vol. 38, no. 2, p. 278. Salt Lake City.

REYNOLDS, S. H.
1913. The vertebrate skeleton. 535 pages. Cambridge, England.

ROMER, A. S.
1955. Vertebrate paleontology. 687 pages, Chicago.

SCHWARTZ, C. W. AND E. R.
1959. The wild mammals of Missouri. 341 pages. Columbia, Missouri.

SIMPSON, G. G.
1945. The principles of classification and a classification of mammals. *American Museum of Natural History, Bulletin*, vol. 85, 350 pages. New York.
1951. Horses. 247 pages. New York.

SISSON, S., AND GROSSMAN, J. D.
1953. Anatomy of the domestic animals. 972 pages. London.

STAINS, H. J.
1959. Use of the calcaneum in studies of taxonomy and food habits. *Journal of Mammalogy*, vol. 40, no. 3, August, pp. 392–401. Salt Lake City.

STEWART, T. D.
1959. Bear paw remains closely resemble human bones. *FBI Law Enforcement Bulletin*, November, pp. 17–21. Washington, D.C.

TAYLOR, W. T., AND WEBER, R. J.
1956. Functional mammalian anatomy. 575 pages. Princeton.

TAYLOR, W. W.
1957. The identication of non-artifactual materials. *National Academy of Sciences, National Research Council, Publication* 565, 64 pages. Washington, D.C.

WARREN, E. R.
1910. The mammals of Colorado. 300 pages. New York.

WHITE, J. A., AND DOWNS, T.
1961. A new Geomys from the Vallecito Creek Pleistocene of California. *Los Angeles County Museum, Contributions to Science*, no. 42, June, 34 pages. Los Angeles.